D0852566

RECEIVED
OCT 2019
By

REWILD YOURSELF

ALSO BY SIMON BARNES

The Meaning of Birds

REWILD YOURSELF

MAKING NATURE MORE VISIBLE IN OUR LIVES

SIMON BARNES

PEGASUS BOOKS
NEW YORK LONDON

REWILD YOURSELF

Pegasus Books Ltd.
148 West 37th Street, 13th Floor
New York, NY 10018

First Pegasus Books hardcover edition October 2019

ISBN: 978-1-64313-216-7

10 9 8 7 6 5 4 3 2 1

Printed in the United States of America
Distributed by W. W. Norton & Company, Inc.

This one's for Kim – perhaps the
brightest witch of her age

Contents

INITIATION

And so Lucy found herself walking through the wood arm in arm with this strange creature as if they had known one another all their lives.

– *The Lion, the Witch and the Wardrobe*, C. S. Lewis

We're not just losing the wild world. We're forgetting it.

We're no longer noticing it. We've lost the habit of looking and seeing and listening and hearing. We're beginning to think it's not really our business. We're beginning to act as if it's not there any more.

In the course of *The Chronicles of Narnia,* Lucy has to enter the house of the magician and read a spell for making hidden things visible. She knows at once that she's got it right, because colours and pictures start to appear on the pages of the magic book in gold and blue and scarlet. And then, wonder of wonders, Aslan himself, the great lion and high king above all high kings, is also made visible.

'Aslan,' said Lucy, almost a little reproachfully. 'Don't make fun of me. As if anything I could do would make *you* visible.'

'It did,' said Aslan. 'Do you think I wouldn't obey my own rules?'

Lucy's joy when Aslan is revealed before her is the most wonderful and beautiful thing, the most wonderful thing that could ever be. And Lucy's joy – or at least some of it – can be ours.

We too can say the spell for making hidden things

visible. In fact, I am prepared to offer no fewer than twenty-three spells for doing so.

Mammals you never knew existed will enter your world. Birds hidden in the treetops will shed their cloak of anonymity. With a single movement of your hand you can make reptiles appear before you. Butterflies you never saw before will bring joy to every sunny day. Creatures of the darkness will enter the light of your consciousness.

As you take on new techniques and a little new equipment, you will discover new creatures and, with them, areas of yourself that have been dormant. Once put to use, they wake up and start working again. You become wilder in your mind and in your heart.

That's the real magic. You wake up the part of you that slept. It was there all along. It needed only the smallest shake, the gentlest nudge to become part of your waking self. There is wildness in us all, but in most of us it's latent, sleeping, unused. Wild we are in our deeper selves: we are hunter-gatherers in suits and dresses and jeans and T-shirts. We have been civilised – tame – for less than 1 per cent of our existence as a species.

Even in the twenty-first century you can be where the wild things are. These days, non-human life always seems to be just over the horizon, just beyond the threshold of our understanding, just a little bit short of our awareness – but with the smallest alteration all this can change. The lost world can be found: the hidden creatures that share our planet can be brought before us glowing in gold and blue and scarlet.

Once you know the spells, the wild world starts to appear before you. Do you think they wouldn't obey their own rules?

Now you don't see it.

Now you do.

1

The Magic Tree

Everyone in that crowd turned its head, and then everyone drew a long breath of wonder and delight. A little way off, towering over their heads, they saw a tree which certainly had not been there before.

– *The Magician's Nephew*, C. S. Lewis

There is a Magic Tree in Narnia. It grows in *The Magician's Nephew,* which is what we now call a prequel to *The Lion, the Witch and the Wardrobe.* Polly and Digory fetch the apple from which the tree springs, making the journey on the back of a winged horse. The tree protects Narnia from harm for hundreds of years and works further wonders when the children return to England.

The idea of magic trees runs deep in all human cultures. Every year we bring one into our homes, cover it with beautiful things and then sit before it in joy. All trees have some kind of magic about them: huge things that start as an object you can hold between finger and thumb, living things that are also life-givers, offering food and shelter to all comers.

I want to draw your attention to one kind of tree: a tree with a summoning spell – and what it summons is butterflies. Butterflies, more than any other creatures, seem to have been designed to please humans: stunning little fragments of colour that sting not and cause no harm. They are without question bright and beautiful, and to look at butterflies with the tiniest bit more attention is the easiest way in the world to get a little closer to nature.

The tree in question is the buddleia. Not much of a tree in terms of height or girth, not much more than an ambitious bush. But it's rather good at flowering: each summer, around June and July, it produces improbable quantities of big purple cones, flowers that seem more than their fragile stems can bear. These blooms summon butterflies and the butterflies obey the summons in unexpected numbers.

The buddleia is classified as an invasive species. It is native to China and Japan and first appeared in British gardens around 1800. It was first recorded in the wild in 1922. It makes plenty of seeds and they disperse well, and it's so spectacularly unneedy that it will put down roots in a wall. More than anything else, it loves to follow railway lines. Buddleias can damage buildings and have also caused problems on chalk grasslands, so not everybody loves them, but they have an utterly beguiling trick. They attract butterflies like no other tree, like no other flowers.

The buddleia is the bar no butterfly can pass. They come in like absinthe drinkers in an impressionist painting and as they drink, they spell out their own theories of colour and meaning. If you want to find butterflies, all you have to do is find a buddleia in bloom. Gaze at it for a few minutes on a fine day in early summer and you will almost certainly find yourself gazing upon butterflies.

There are plenty of nectar-rich plants, and many that attract butterflies, but the buddleia has the power of summoning butterflies from vast distances and it makes all other plants seem second-best. Here the butterflies drink and bask, spreading their wings out just like the illustrations in the field guide, making it easy for the greenest of novices to make an ID. They are so taken up with the glories of the buddleia that they will more or less ignore your presence. You can walk within a couple of feet of the bush and be unlucky if the butterflies move. The buddleia gives you a unique, almost impertinent intimacy with butterflies.

So far so good. Now all you have to do is to meet the tree halfway. Do that and you will work a spell that diminishes the gap between human and butterfly. And what you do is learn to tell one species of butterfly from another. Do that and you will never look at a butterfly in the same way. That's because every time you see one, you will look a little closer, know a little more and the butterfly will have just a little more meaning.

From that moment on you will see every butterfly more clearly than before. You want to know what species the butterfly is, just as you want to know who is calling at your house. And as you get used to the fact that butterflies come in different species, you start recognising butterflies whether you want to or not. That's not a butterfly – that's a peacock.

You have programmed part of your brain with butterfly images: with the images of different kinds of butterflies. Once you have sorted out a few species to your satisfaction, your brain will respond to butterflies in a different way. There will be knowledge, familiarity and, above all, connection. It may be one-way, at least in any obvious sense, but you will find yourself connected to butterflies as never before.

As Danny the drug-dealer says in the film *Withnail and I*: 'You have done something to your brain. You have made it high.'

And that is the way you rewild yourself: you do something to your brain. This book, and every spell in it, is about doing something to your brain. You can make it wild.

The buddleia bush is the summoning spell that brings the butterflies to you. You complete the charm by learning five names – and then find that they lead to a still-more-magical sixth. All butterflies are equal, but some butterflies are more equal than others. We all love the idea of the Special One.

There are fifty-nine species of butterfly regularly seen in Britain, so it's easier than with birds, when you have to cope with more than three hundred. With birds, you can increase your enjoyment hugely by learning the names of forty or fifty of them. But right now I'm talking about butterflies, and the fact is that you can increase your pleasure in butterflies, and therefore of every summer's day you ever live through, by learning the names of the five butterflies that you will routinely find on a buddleia. The classic garden butterflies; the butterflies who pass by when you're having a nice drink outside on a warm evening. And then the magical sixth . . .

The first is the small white, often disparagingly known as the cabbage white; its caterpillars have a taste for brassicas and, for my money, they're welcome to them. But that's by the by. The point is that if you see a small white butterfly on a buddleia, the chances are it's a small white. Not all of life, and not all of wildlife, is as simple as this, so it's worth celebrating.

And then, as an incentive to look closer, you may also find a large white. This is indeed large and white, but scale can be tricky in all forms of observation. The large white – also fond of brassicas – has strong black corners to the upper wings; the females also have two black wing-spots.

There are other white butterflies to confuse things but, right now, I suggest you pay them no mind. Keep it simple. Just start looking at white butterflies and separating the large from the small, and you will already find your eyes and mind are turning feral on you. You're waking up. The process of personal rewilding is beginning.

Then there are three colourful butterflies, all equally keen on buddleias. Let's start with the small tortoiseshell. This was the breakthrough butterfly for me, and I hope it will be for you. I spent too much of my life thinking that butterflies weren't really my business and that they were all, well, just butterflies. I started looking at butterflies mainly because it seemed really rather feeble not to know one from another. The second reason is that the peak flying months for butterflies – the midsummer – are the quietest for birds, as they're involved in raising a family rather than being noticed.

It's a relatively recent trend: birders turn to butterflies in the summer, often to dragonflies as well, which are a good deal trickier. So I learned the buddleia boys and for ever after I have looked at butterflies and seen them where I would never have seen them before. That's not because they weren't there: it's because I hadn't looked. I hadn't known how to see. I hadn't done something to my brain.

The small tortoiseshell is a warm orange but the leading edges of the wings carry a bar of alternating black and yellow. They are tortoiseshell in the way that tortoiseshell cats are, or sometimes the frames of glasses: that same black and orange mixture. (A butterfly has four wings, two on each side, so close together they often look like one.)

Now look closely as a small tortoiseshell spreads his wings before you and holds still: you will see that the trailing edges of the hind-wings have the most exquisite pattern of blue dots. When I started looking at butterflies, I had never seen it before, never noticed the blue at all. How absurd it is: a miracle of colour that regularly comes into back gardens and parks, not to mention railways lines and car parks, and I hadn't seen it at all. I had looked at it many times, but I hadn't seen it. My brain was insufficiently wild. With the smallest amount of training you can see it every summer's day.

The next butterfly to look for is the peacock. On a rich red background, four mad staring eyes. That's why it's called a peacock: a peacock's tail is similarly full of eyes. The peacock's eyes are for showing off to the females; a butterfly's eyes are for startling predators: for giving out the lying information that this is not a tasty insect but part of a large and possibly dangerous beast. You may say, 'Well, it doesn't fool me', but that's because you're looking with a human's two-eyed perspective. An insect-eating bird has eyes set on either side of its head, so its three-dimensional perception is not acute. It is also looking around for predators all the time: a quick impression of a fierce eye and it's off. Better safe than eaten.

Mimicry in butterflies (and many other beasts) is fascinating, bewildering and sometimes close to hallucinogenic. You can find stunning examples of this in the rainforest – and in anyone's back garden. The great spectacles and mysteries of life are not restricted to distant lands: they are within your compass – your garden, your local park, your eye, your brain. That is what rewilding yourself is all about.

The final one of the buddleia five is familiar to many of us, and a good few of us even know its name. That's the red admiral: shining black wings picked out with white and red, a handsome beast and about as burly as a butterfly can get. They are strong fliers and indefatigable migrants. They also hang about till later in the year than most butterflies: long after your buddleia has gone to seed you can find red admirals on ivy flowers. Not everyone notices that ivy does flowers: red admirals do, and they will draw your attention to these immensely discreet blossoms; a great source of nectar when most of the other flowers have gone. Thus the red admirals refuel on their way south to the mainland of Europe.

So let us move on to the magical sixth: the Special One. Not there every time, not by any means. Not there every year, for that matter – but some years they're here in abundance and you can rejoice, and wonder if they're going to break the buddleia with their mass, as it seemed they might do in the great year of 2009. These are painted ladies. They too are migrants and they fly up from Morocco to be with us. They vary a little, but mostly they're pale orange, with the outside corners of the wings black, and white markings within the back.

They don't come all the way from Morocco in one go: they will pause and breed and die and then the next generation powers northwards. It was once thought that they came to Britain to produce a doomed generation, one unable to survive the British winter: but not so. It's been discovered that they make a southerly migration, often flying very high, where they have been tracked by radar. There's a moral there: never underestimate nature. Never patronise nature. Nature is better than we know, probably better than we'll ever know.

Now pause to savour that. These are butterflies: animals that we associate more than anything else with fragility and purposelessness. 'Who breaks a butterfly upon a wheel?' asked Alexander Pope. A butterfly mind belongs to someone who can't hold an idea in his head for longer than a few minutes.

But painted ladies are powerful creatures that cross the Channel on a routine basis. They are an emblem of strength: strength of body, strength of purpose. And there they are in the back garden, drinking deep of the buddleia: just another miracle of everyday life.

So take these six butterflies into the summer with you, and your summer will be richer. As you do so, you will find yourself looking at other butterflies – and that's when you realise that your brain is getting wilder. You find yourself taking delight in the dashing flight of a male orange-tip – small and white but with orange wing-tips – or in the little jewel of a common blue, or in the gorgeous flash of a brimstone. Brimstones are the colour of butter – the best butter – and it's reckoned that they are the butterflies that put butter into the name of butterflies.

So you can enter another country – the wild country – not through a wardrobe but by means of a Magic Tree. Enter, then, with joy. And after that, you can turn your mind to another spell.

2

Magic Trousers

It's the wrong trousers, Gromit – and they've gone wrong!

– Wallace in *The Wrong Trousers*

These are the trousers that will change your life. They will make fundamental alterations to your attitude to the world. They are the nearest you will ever get to a superpower. And they are waterproof trousers.

You no doubt possess a waterproof top, and perhaps a reasonably hearty one that can be worn without embarrassment along a country path. But waterproof trousers are a serious escalation. They are an expression of comparatively serious commitment to the idea of being outdoors: an acceptance of the fact that the outdoor life will involve you in inclement weather.

We are used to being slaves to the weather. The weather tells us if we should be indoors or outdoors. The weather tells us how much we are to enjoy ourselves. The weather orders us home or into shelter. The weather can take all the pleasure from the outdoors. However, if you restrict yourself to nice days, you spend a lot of time indoors. A lot of time not being wild.

I got serious about waterproof trousers when I walked the Cornish cliff path with my father many years ago. We had long talked about this project, but there came a time when my father suggested that the time to do it was now, 'before double incontinence sets in'. So we made our plans, and in doing so realised that the task of walking to your next night's accommodation was likely to involve walking through bad weather.

'No such thing as bad weather, only unsuitable clothing,' said Alan Wainwright. So we carried waterproof trousers and, when the rain came, we put them on. When the rain cleared, we generally took them off again. And when the rain came again, we put them back on again. 'Up and down,' my father said. 'Like a tart's drawers in Navy Week.' I now possess a pair of waterproof trousers slashed to the thigh, like the Chinese female garment known as *cheongsam*. That makes them easy to put on over boots, and mostly gets rid of the counterproductive need to sit down (on wet grass) to put your waterproof trousers on. You then zip them up – or rather down – from thigh to ankle.

When you have waterproof trousers on, you are master of the universe, lord of the weather, and you are in possession of a visa that allows you to enter unknown lands.

I was on Fair Isle: a tiny dot of an island with a permanent population of around fifty; it lies between Orkney and Shetland and is one of those truly unforgettable places. I was researching a story about the islanders' courageous and successful fight to establish a Marine Protected Area around their coast. This involved a series of interviews with the influential people of the island, and it was all rather intense, as indeed it should have been. This was serious business, after all.

But as I left the house of one of the island elders I realised that I had a couple of hours free, and no one to meet. It was a chance to see the island – to get the hang of the island – on my own. The only trouble was that it was raining. It was a special kind of rain, one you don't find every day or everywhere: it was only just about capable of falling. It was a little heavier than mizzle and, of course, perfectly soaking. It was like being inside a cloud.

So I put on my waterproof top and, better still, my waterproof bottom and set off to savour the island in what seemed ideal conditions. I walked for a bit, and that was fine, but it felt too . . . too purposeful. I needed to be inconsequent. Now anyone can be inconsequent on a sunny day, it's one of the reasons we welcome a day of sunshine. We can walk at our ease and sit at our ease and take the day on its own terms.

But, dressed as I was, I could be inconsequent in the midst of weather that would otherwise have sucked all pleasure from the day. I clambered down onto a beach and – pause to savour this miracle – I sat down. And, dry-bummed, I remained sitting. The islanders had been telling me how they loved the island, now I had an opportunity to see why. They had told me the island mattered: now I could let the island itself tell me the same thing. They told me this was a special place, that they were all deeply privileged: now I could share that privilege. And all because of my trousers.

The professional concerns of gathering material and adhering to a schedule and making my connections back to Shetland the following day, all these things could be set aside. I sat there on wet grass, my back resting against wet rocks, and the air itself was wringing wet. Before me the sea heaved and shifted in its unquiet way, drawing the eye as any body of water always does. I looked out at the seabirds, fulmars, who are related to albatrosses, and kittiwakes, the true sea-lovers among British gulls. And as I watched, a head appeared from the water, stared at me gravely, and then disappeared.

I wasn't imagining this: the head was real and so was the interest in me. Not hostile, just keeping an eye. Perfectly at ease with me so long as I did nothing stupid.

And then another head rose, gazed, sank: and another, and another. Sometimes there were twenty heads in view at once, sometimes just one or two. These were seals, of course, Atlantic grey seals – big, imposing carnivores that have no very good reason to trust humans. But in their own shifting domain of the bay, they were confident that nothing would go wrong, and so I sat on the wet beach while they sat in the wetter water and we did so together for maybe an hour ... and by the time I left the shore I felt as if I had lived on the island all my life and would live there till I died. Of course I knew that this was an illusion, but this deeper sense of involvement with the place came from sitting there, still and quiet, in conditions that would have driven me to shelter had it not been for the miraculous trousers that enclosed me and kept me safe.

And do you know something? Just before I got up to leave, one of the seals raised his voice in song.

That, then, is something that waterproof trousers can do for you. What these trousers do is break down some of the barriers between being inside and being outside. Suddenly you are less cut off from the wild world than you were before: you can step out and be part of it, heedless of showers, fearless of downpours.

It's not just about keeping dry: it's about changing your notion of your place in the world. You are free to step out in a downpour – and not just survive. You can actively enjoy it. You learn that downpours have their place in the world, not as something to view from the far side of a pane of glass, but as something to walk through with confidence – and, should you wish, to sit in comfort and to enjoy.

Sometimes you will see some nice wildlife by doing so. That's a bonus. What you're really doing is breaking down one of the restrictions that we have imposed on ourselves: the idea that we can only go out into the weather to enjoy ourselves when the weather is nice. Because it's not about being nice: it's about being wild. The right trousers will release in you a level of wildness you didn't know you possessed. The trousers cut you off from the wet – and by doing so they connect you more deeply with the wild.

3

How to Enter a New World

'This must be a simply enormous wardrobe!' thought Lucy, going still further in and pushing the soft folds of the coats aside to make room for her.

– *The Lion, the Witch and the Wardrobe*, C. S. Lewis

It's one of the greatest coups in the history of storytelling: the moment when Lucy enters a wardrobe during a game of hide and seek and finds herself in a new and magical country. It has immense meaning and resonance for us all.

First, it's a perfect recapitulation of the imagination games of childhood, when the least promising items of domestic furniture assume fantastical and magical properties: beneath the kitchen table is an immense system of mysterious caves, the arm of the sofa is a horse, and then a fort to defend from a thousand enemies.

Second, this idea of a magical new country is a longing that remains with us in adulthood: how many times in our life do we wish to escape from tedium into adventure – from restriction to freedom, from a circumscribed life to a world of infinite possibility – and all in the space of half a dozen strides? Such countries include Provence, the Costa Brava, Barbados, a desert island, the British countryside . . .

And third, the truth is that we can do exactly that. It's just a matter of knowing how.

We can make a magical transition from one kind of place to a completely different kind of place and do so, if not quite instantaneously, then certainly within astonishingly few minutes. We can move from the familiar to the deeply improbable in less than half an hour – and all for the cost of a few quid. Here's a fiver: take me to another world! Right away – sit down and make yourself comfortable and let's go.

Birds are often the clue. They tell us when we are in a pleasant city park, they tell us when we are sitting by a lake, they tell us when it's summer and when it's winter. They tell us, often loudly, that we have moved to a different place, a different world. Like the seaside. Proper birders scorn the term seagulls, so let's use it wildly and recklessly. Mostly, seagulls are any one of five different species of gull that we meet when we go to the seaside, and their voices constantly remind us that we are in a new and unfamiliar place. Listen to *Desert Island Discs*: the display-calls of herring-gulls tell us that we are now bound to the theme of the shore.

But if we think that these five seagulls are what the sea is all about, we are in serious error. Seagulls would be better known as shoregulls. They are birds of the interface between sea and land: half-and-halfers, huggers of coasts, birds that like to keep the land in sight. They are essentially part of the same world as ourselves: but what they do is offer a raucous, screaming invitation to step into another world: a world for which their loud cries are but the gateway.

Perhaps you have thumbed through a bird book and said, 'This bird is not for me. Too rare, too difficult, beyond my scope.' Or perhaps you have seen pictures of birds and other living things and reflected that these creatures are for ever out of reach. The fact is, you can make some of these become visible in the simplest way in the world: by paying your fiver and stepping on board.

For years, I thought a gannet was one of those utterly impossible birds. I knew it was rare, for, after all, when did I ever see one? This was a bird far beyond my hopes and my ambitions. Then, one year in Cornwall, on our annual family holiday, two or three of them appeared. They were fishing just off the cliffs as we were all making a pleasant walk. They were doing all the things that gannets are supposed to do: flying on six-foot wings and diving head-first from fifty feet, folding those vast wings back beyond the end of the tail until the birds assumed the silhouette of an immense dart chucked at the treble-20. It was a like a casual visitation from a flight of angels.

I thought this was a freak, an impossible slice of luck. Then I learned the secret. And discovered that in many places, seeing gannets is the easiest thing in the world. You just have to go from our world to their world. And their world is the sea. Not the shore: the open ocean. So we have to make a small effort to enter that world.

There is scarcely a seaside resort that doesn't offer a nice boat trip: usually lasting for an undemanding hour, though you can find more adventurous operations with a bit of research. You can also work the same spell on a passenger ferry: just make sure you are out on deck, rather than hunkered down under cover. The point is that the easiest and most obvious sea trip takes you through the wardrobe into that unfamiliar world: and you can get a head-spinning taste of what that world is really like.

In Cornwall and many other places, you will learn that gannets are easy to see: all you have to do is to get away from land and look out towards sea. Once you are a couple of miles away from land, you are highly likely to find them.

It really is as easy as that: one simple transition and the impossible is commonplace. The sea truly is another world, and one we seldom visit in the general way of things. But when we do so, it becomes at once bewildering, thrilling and intoxicating.

And I say this as no great lover of boats. I have never once stepped on board a seagoing vessel without trepidation. Once we get going I stare resolutely beyond the boat: lose eye contact with the horizon and the horrors of sickness will surely descend. I remember issuing an order to a party: no jokes! Not even the faintest humorous reference to seasickness is permitted. Start that nonsense and we'll all go. I am no hearty seadog, but I am in love with the fabulous transition into another world that the shortest of sea trips will give you. (I should add here that I was once given an infallible cure for seasickness: sit under a tree.)

I remember seeing hundreds of shearwaters, all of them

moving urgently from one fishing ground to another, for out there the world is forever changing and the great shoals do not stay long in the same place. You can often see small birds that fly close to the surface on whirring wings with all the grace of bath toys powered by an elastic band. These are the auks: razorbills, guillemots and puffins.

There are also possibilities of other creatures that prefer the life of the open sea. Where else in Britain could you have a close encounter with a carnivore more than ten feet long and weighting more than forty-five stone? A bull Atlantic grey seal meets that description and can be seen on many a boat trip: sometimes hauled out onto offshore islands or inaccessible headlands, sometimes swimming by the boat, at other times dozing in the water, upright like a vast bottle, nose occasionally breaking the surface to breathe.

I have also seen dolphins, porpoises and, once, five basking sharks all together, and the smallest of them ten feet in length. One of the most bizarre things I have ever seen was a few miles off the British coast: a sunfish, a creature that can weigh as much as a ton; they love to come to the surface on calm, sunny days and bask.

All these wonders are there: underneath or on the surface or above that heaving great wet mass. And it's here that I must take a slightly moral tone, as Aslan does with Lucy after she has taken advantage of a spell to eavesdrop on her friend.

Some trips can offer a near-certainty of easy wonders, and I'll get to some of them in a later chapter. With other trips, you have to take your chance and risk disappointment. But here's a truth and it's also part of the spell: it's not just about seeing fantastic beasts. The most important part of the experience is being where the wild things are. You open yourself up to the possibility of wonders, and that in itself is a great thing. You don't get all cast down when you don't see something you fancied: half the reason for being out there is the search. The knowledge that you are breathing the same air as dolphins is itself a fine thing.

You're not going to the zoo. The money you have paid is not a contract with the gannets or the dolphins to become visible: it merely takes you through the back of the wardrobe into the unfamiliar land in which such creatures have their being. Just being in it yourself is privilege: just being there is a profound experience of wildness.

Most of the world's surface is ocean. For most of us it's as alien as the surface of the moon, and yet we think we have a sort of understanding of it because – as with the moon – we can see it with our feet on solid earth. But as soon as we make a journey out – even onto the nearest edge – we start to understand that it is, indeed, a different world, very different to the one we inhabit from day to day, and that life continues there in many strange and unfamiliar forms, and yet we can intrude and share that life for a few fleeting moments.

4

The Snake-Charming Spell

As the snake slid swiftly past him, Harry could have sworn a low, hissing voice said: 'Brazil here I come ... Thanksss, amigo.'

– *Harry Potter and the Philosopher's Stone*, J. K. Rowling

There are snakes six feet long slithering across the British countryside. Or, to be more technical, travelling by means of lateral undulation and rectilinear locomotion: the two most significant snaky gaits. But have you ever seen a British snake that size? They're out there, as long as a man is tall, and yet they're superbly gifted at keeping out of human sight.

A snake, perhaps more than any other creature, is built for discretion. Long, thin and legless, they are brilliant at dark, low places, at narrow spots where no one comes looking. They are at home in spaces that seem too small for anything bigger than a mouse: like a living reverse TARDIS, they seem capable of hiding in places smaller than themselves. They hunt by stealth, move by stealth, live by stealth.

I have seen six-footers in Britain just twice. One was freshly dead: having made a rare and disastrous foray out of concealment and onto a road. But the other was in her pomp – and I say 'her' because if she's that big, she can only be a female grass snake. She swam across a large pond with immense confidence – observe those lashing lateral undulations, modified for aquatic use – entered the vegetation on the far side and ... vanished. One, two, three, and where's your reptile? How can something that big just disappear?

Because she's a snake. It's what snakes do.

I have also seen venomous snakes in Britain in hefty numbers – a dozen at once – just twenty yards from a path on which dogs and their walkers made their noisy, cheerful progress through a forest, quite unaware that all these potentially lethal beasts were so close at hand. I saw them because I'd been shown: alone, I'd have seen only piles of brash and bracken and not the bracken-patterned bodies subtly shifting . . .

It seemed on the face of it quite ludicrously dangerous: we have a strong response to snakes, part taught and part innate. The idea of a dozen venomous snakes a cricket-pitch away from the thin-socked ankles of innocent walkers gives pause for thought. (To be accurate, poison is stuff that harms you when you ingest it; venom does so when it's injected. You can drink snake venom without harm – unless you have a cut in your mouth. So although some snakes are venomous, they're never poisonous.)

And yet none of those dog-walkers, and for that matter none of those dogs, was aware of the secret slithering and genuinely venomous presence of those fine adders, newly wakened from their winter slumbers and ready to embrace the sun and the warmth of a fine spring day. Such a day puts a spring into everybody's stride, legs or no legs. A spring into your lateral undulations, perhaps.

So yes, there are reptiles in Britain, successfully living their lives while staying almost entirely out of sight of human beings. They're a classic example of the invisible life of our island. And while it's a pleasure to think of the legions of secret snakes forever just out of sight, it's also fun to see them every now and then.

The spell for making hidden reptiles visible is improbably simple. It requires all the skill you need to throw your McDonalds wrapper into the hedge.

You need a piece of tin.

Or, to be more accurate, a piece of corrugated metal. Such a thing can be your snake-charmer: magicking your snake and persuading it to come to you. So much easier than you going to the snake, after all.

The metal needs to be a square metre or less, down to about half a metre. Or to put that another way, each side needs to be half to a quarter of the length of that gorgeous female grass snake. All you have to do is put it down in a likely place.

Then you come back on a likely day and lift it up. And most times you find ... well, exactly what you'd expect: absolutely nothing. But every now and then – and you can quadruple your chances if you pick the right time of day and year and the right weather – you lift it up and beneath it you find a reptile.

Bloody hell!

That's what you'll say. At the very least.

My place in Norfolk is part of a damp and watery landscape. That means that when I lift my tins, if I find anything I find a grass snake. Never yet a whopper of the size of the great pond-crosser, but maybe a dozen times a year I find these perfect and beautiful things, a couple of feet long and with a yellow collar so bright you'd think it might hurt your eyes.

And it makes me jump.

Bloody hell!

I know they can't harm me, and I'm not phobic about snakes – but that dramatic revelation is still ever so slightly disturbing. You never expect to see a snake, not anywhere – not in the African savannah, not in Asian jungles, and certainly not in Britain. A snake always looks alarmingly unlikely: as if the real world had been taken over by creatures of our fevered imagination. Seeing a snake in Britain seems almost as unlikely as seeing the great sea-serpent from the Isle of Wight ferry.

Reptiles creep beneath these tins because they can find warmth under there. And reptiles need warmth. We call reptiles cold-blooded, but that's a rather dodgy term. They like their blood good and warm, but they can't generate their own heat. They have to borrow it. It's a different strategy to our own, and always there's a trade-off. Warm-blooded creatures, like us mammals, generate our own heat, and for that we need fuel. That is to say, a lot of good-quality food. We need to eat all the time because all the time we're burning the stuff up in our central heating.

But a reptile is built for economy. It needs to eat only occasionally. The massive crocodiles on the rivers of the Masai Mara and the Serengeti eat only once a year, but then prodigiously: they wait until the wildebeests and zebras cross the rivers and then they pounce.

This kind of economy is a very effective strategy, but the payback is that when the weather is cold, a reptile can't do very much at all. In winter they shut down entirely.

Hence the tin. If a reptile gets under a tin, in the right weather – like a spring day full of unseasonal warmth – it can find a lot more free heat than is available outside. So it gets underneath, then it gets warm, then it gets active.

A sunny morning in March gives you a great opportunity for success with your tin. Mornings and late afternoon are the best times as the season progresses. Any time between March and October gives you a chance: in late summer you might find the year's hatchlings as they begin to reach a half-decent size.

It's all about weather. Warm days in spring and autumn are your best bet, and April and May are probably your best months. But it's always worth raising your tin every time you pass, on the off-chance.

So where do you put them? Forgotten, quiet places, away from well-trodden paths. All kinds of unexpected spots might be profitable. Gardens can be productive.

Disused corners of anywhere fairly wild are possibilities: allotments, golf courses, railway and roadside verges.

Anyone can buy a chunk of corrugated metal easily enough. You can even get them specifically designed for the job, for not very much money at all: search online for reptile refugia.

Britain is a bit too cold for most reptiles, which is why we only have six species; studying British reptiles is a bit easier than being an entomologist, who has a choice of 20,000 British species. Three of these British reptiles are snakes: grass snake, adder and smooth snake; only the adder is venomous. Then there's the common lizard and the sand lizard. The last is the traditional teaser: the slow worm, which had no legs and is a reptile, but it's not a snake: it's a lizard that happens to have no legs. Slow worms can shed their tails and blink their eyes; a snake can do neither. And they're lovely things, especially when caught in sunlight, gleaming like living bronze.

So what do you do when you've lifted your tin and seen a lovely reptile? Do what I do: lower your tin and walk on, slightly richer than you were before. You need a licence to handle sand lizards and smooth snakes, they being rare. But you don't need to handle any of them: better not, in my view.

Anyone who needs a health-and-safety warning about messing about with adders is pretty foolish to begin with. There's no need to get hysterical: no one in Britain has died from an adder bite for twenty years, but being bitten can still be a highly unpleasant experience. More than 90 per cent of all snake bites come to people handling snakes; a helpful hint, I find.

You can move up to a greater level of seriousness and send your observations to Froglife, a conservation charity for amphibians and reptiles. But you can also just enjoy the moment of magic when you cause an invisible reptile to become visible: to appear briefly in glory and wonder. They're out there, but you'll hardly ever see one – unless you happen to know the spell.

5

A Magical Journey to the Turd World

Feel! Press! Crushed! Sulphur dung of lions!
Young! Young!

– *Ulysses*, James Joyce

I was discussing the day's programme at the visitor centre and administrative headquarters of Badlands National Park in South Dakota when a lovely girl rushed into the room.

'Oh God!' she apologised. 'I can't find my poops.'

Then she realised there was a stranger in the room. 'I'm sorry,' she said. 'You have no idea what I'm talking about, do you?'

'But I do,' I said. 'And I have something of interest to you right here.'

I took my phone from my pocket and showed her some photos. Three photographs of excrement. All different kinds of excrement – and how wonderful is that? Dakota – for that was her splendidly appropriate name – gazed at them with genuine fascination. 'Well, I know this lot is Chinese water deer because that's the deer we get round my place,' I said. 'They're not vastly different to Reeves's muntjac, but they don't come down onto the marsh.'

The story has a happy ending. Dakota found her poops, and I watched her give a presentation on the mammals of Badlands National Park to local schoolchildren. She was absolutely first class: full of that combination of love of subject with well-organised knowledge that marks all the best teachers. Afterwards I was able to examine her poops close up – well, you can imagine the sense of privilege. Bison, pronghorn, coyote, black-tailed prairie dog – not dogs at all but highly social rodents with remarkable communication skills – and Dakota's speciality: the rare and reintroduced black-footed ferret.

You hardly ever see a black-footed ferret, not even if you know exactly where to look. But you can find their droppings, and thereby know that they are thriving. You can even, if you persist, find out what they're thriving *on*. If you know how to look, if you know where to look, if you know how to identify the turds, you make these hidden mammals visible.

Perhaps you have wondered, dear readers, why so many people go bird-watching and not mammal-watching, even though we're mammals ourselves. Why don't we go out watching our own kind?

We don't because, on the whole, they're very difficult to see. Most mammals tend to be active at night, or at least around dawn. They are very good at concealment. Birds – like humans – are creatures of colour and sound. Most mammals live in a sensory world in which smell is hugely – to us incomprehensibly – important. We primates don't: like birds, we're better at seeing and hearing. Most of our fellow mammals inhabit a sensory world beyond our imagination.

It can also be hard for wild mammals to set eyes on others of their kind. But they can leave messages for each other in scent. For them, a turd is not just a waste product, to be discarded and forgotten: it's a thing of immense significance and meaning. A dog's fascination with the droppings of another is not a foul and disgusting habit: it's the newspaper, the internet, the Christmas round robin. That's why so many mammals leave their droppings in a conspicuous place. The animals themselves are hidden but what they leave behind is a great olfactory trumpeting of their presence.

It follows, then, that an awareness of turds is a spell that will bring hidden populations of our fellow mammals into human awareness. Especially when we add to this a very little knowledge.

So there I was on my knees in an area of marshland on Norfolk, this time in the company of a lady called Helen. And we were both filled with delight, for we had found a turd. Who could fail to rejoice? For it showed beyond all possible doubt that otters were living here, using this system of dykes, using the same path that we were walking on.

Otters spend a good deal of time on their own, though they can be wildly social on occasions. But, like most mammals, even when they are alone they live their lives in the context of others of their kind. Though they seldom see each other, they like – they need – to know all about each other: whether the big male is still around, who that younger male is that turns up from time to time, whether that female has come into oestrus. And so they leave messages – fridge notes, if you like – and these take the form of turds. Thrilling, information-rich, and usually laid in some prominent position. Drop under a bridge in a likely-looking stream and if you can find a raised spot under there, you may well find a turd; spraint is the correct term here. It's information and it's also a territorial claim: so if you're that young otter, watch your step, because this little offering tells you that the big male is still around and in his prime.

This is the way otters live and understand the world. It's also a way that we humans can begin to understand otters – or at least get a clue about their existence.

What's required, then, is a subtle mental adjustment. What was once a thing of mild disgust must become an object of fascination: something that brings hidden creatures into your circle of understanding.

A garden lawn is a good place to start. Say, beneath a bird table. You might find a turd about the size of your little finger, maybe black and shiny, and not too horribly stinky. It may well be round at each end, rather than tapered. If you like, you can take your involvement a little further and tease the thing open with a twig. You will probably find a lot of shiny little bits of insect – and what you have found is a hedgehog dropping. No: what you have found is a hedgehog. You have revealed a nocturnal visitor you wouldn't otherwise have known about, bustling about his own business and using your garden to help him make a living.

If you have a smartphone of some kind, then you can take a photograph. You can use this to impress your next date, but almost as importantly you can also use it for reference and look it up. You'll be surprised how many helpful pictures of turds you can find on the internet. Here's a hint: take a familiar object from your pocket – a coin, say, or a penknife – and include that in the picture. That'll give you an idea of scale when you come to look at it.

You can find deer droppings in clusters of smallish pellets, like grapes, and it's always good to know that they're about. Rabbit droppings are smaller and rounder: hares' are a little larger, as you'd expect. If you look between the stems of reeds and grasses at the edge of a waterway, you might find the tiny rice-grain-like stuff left behind by water voles.

All the same, I accept that rejoicing over a turd is not always easy. In our sanitised world we do our very best to keep our natural functions as far from us and from other people as we possibly can. We also have a very strong and ancient revulsion to the smell of skatole. We humans mostly smell in black and white, but certain smells blow us away. Skatole is found in excrement, most particularly that of carnivores. The droppings themselves can have a high concentration of dangerous bacteria, and our strong natural repugnance is an ancient warning signal and survival mechanism – keep away! It's not just disgusting, it's dangerous – but it's our disgust that keeps us away. (Fascinating fact: skatole is also found in low concentrations in orange blossom and jasmine.)

All that explains why dog shit is so particularly shitty. But, if we overcome our natural disgust, we can learn thrilling things about our fellow mammals. A dog turd in a place where no dog goes is likely to come from a fox: black, twisted, pointed at one end, often (unlike the droppings of any domestic dog) containing fur and feathers, sometimes shiny with beetles, in season rich with berries.

Turd awareness enriches your life. Turds are best pursued in areas that aren't ruthlessly criss-crossed by dog-walkers and their accompanying turd-manufacturers, but it's always worth keeping your eye open for something less predictable, like the scattering of pellets that tells you that there are deer here, living their secret lives, keeping out of everybody's sight – everybody's sight but yours.

6

Darkness Visible

On the bat's back I do fly
After summer, merrily . . .

– *The Tempest,* William Shakespeare

We humans are creatures of the light. We hate the darkness and everything about it. We equate darkness with evil: in *Star Wars,* the bad characters go over to the Dark Side; in *Harry Potter,* the good wizards struggle against the Dark Lord; people worship the devil in a Black Mass.

Sight is everything to us, or almost everything. That is why the darkness cripples us: leaves us hiding or seeking the light. Unlike most mammals, we primates lack a tapetum: the reflective bit behind the eye that allows its possessors to get double value from the light they collect at night. That's why human eyes don't shine in the dark when you shine a torch on us.

We are adapted for the daylight, and only for the daylight; even low-light situations have us struggling. As a result, we have the deepest distrust of animals who are at home in the darkness, and we load them with all kinds of sinister meanings: bats, owls and toads are associated with witchcraft. We are cut off from such creatures: we inhabit opposite ends of the 24-hour cycle. We call these twenty-four hours a 'day', and we give the same name to the sunlit parts of the cycle. What matters to us, then, are the hours when we can see.

Occasionally, around sunset in the warmer months, we

might catch a glimpse of a bat. It's here – it's there – it's gone! Fleeting, fluttering shadows: was it really there at all? And while that's all very cheering, in a life-teeming-all-around-us sort of way, bats are seldom animals we concentrate on. Which is a shame, because there are seventeen species of bats breeding in this country: getting on for a quarter of all our non-marine mammal species are bats.

Owls are easier than bats, because they make a good deal of noise. They are very keen on communication and they like to know where everybody else is, and they do so by calling, which makes sense for a nocturnal creature with no need to hide from a predator. But bats are silent: just wispy shadows at the tail of our eyes.

Silent to us, that is. Of course, we know that they make sounds. We've read about it, we've been told about it. They make their way securely even through the total darkness of a cave, they locate prey species even on the blackest of nights, and they do so by sonar: by listening to the echoes of their own voices. This is one of the classic wonders of nature, and it seems particularly wondrous to us because we can't begin to imagine what it must be like to be at home in the dark.

Now, if bats find their way through the night by means of sonar, we should be able to hear the noises they make as they do so. But, of course, we don't – though children and some young people can sometimes pick up the odd murmur. Bat murmurs are a privilege of youth: we tend to lose the higher sounds as we get older. I once had the infinitely frustrating experience of listening to the thin, sweet, high song of a firecrest in a conifer and not one of the half-dozen people I was with could hear it. It would have been a new record for the site.

But bats' sonar is beyond more or less all of us. We tend to call such inaudible noises 'ultrasound' as if they were beyond sound itself, but that's a human-centred way of understanding it. The bats' voices are not beyond sound: just beyond the listening equipment we humans are born with. Ours is not the only sensory world, and the more we look beyond our own species, the more we become aware of that. (Most mammals see only in two colours; we primates mostly see in three; many bird species see in four, so their visible world is very different from out own: perhaps incomprehensibly splendid. We can't distinguish between male and female blue tits by sight: blue tits can because the male possesses an ultraviolet crest. Such a bird would find our three-coloured televisions incomprehensibly drab.)

You can bring bats into your world by getting up half an hour before dawn or – a trifle easier for most of us – by getting out to a likely place at sunset. Don't just look up: look out. Look low over water, look all round: bats are astonishingly mobile and impossibly agile. All this in the warmer months, of course; when there are no (or very few) insects on the wing, bats hibernate. All British bats are insect-eaters.

You can find bats in most habitats – yes, even cities.

Woodlands, parks, fields, back gardens: all are possible places for a bat. Bats can fly – the only mammals capable of powered flight – and therefore distance is less of a problem for them than for the rest of us. A sunset excursion will, with reasonable luck, bring you a glimpse of a vanishing shadow: a thrilling realisation of life going on at the fringes of our own senses and our own realisation.

But you can do better than this. You can acquire a bat detector. This is a device that listens to ultrasonic sounds and, in a flash, converts them into sounds within our range of hearing. I won't go into details about the physics, largely because I don't really understand them, even though I got physics O Level, one of the more unlikely triumphs of my life. Many of us find it hard to concentrate when we see something like 110 KHz written down: our brains make a rapid decision that this is not for us and force our eyes to skim on past.

So let's skip the how-it-works stuff and concentrate on what it does. You switch it on and it hisses: you adjust the volume to a level at which you are convinced that it's still working, but you're not deafened by the hiss. After that, you look to the second wheel and select the frequency you want to listen on. The dial will tell you how many KHz: set it for around 45 and listen out as you make your sunset pilgrimage. No doubt you'll be walking up and down and looking for promising spots.

And then – bang! – it happens. A sort of bongo-drum

solo that lasts for two or three seconds. A frantic pattering. And then nothing. And you've seen nothing. But that was a bat, echolocating its way through the gathering darkness.

'Gliding and fluttering back and forth, she shouts her torch of sound among the trees, listening for her supper,' wrote Nicola Davis in her wonderful children's book, *Bat Loves the Night*. Now you can catch the beam of that torch and hear it with your magic ear. Then, almost at once, you want to hear it again. You can stop and see if the bat comes back (they will frequently follow the same flying pattern if it seems worth their while) or you can walk on, aiming your hissing box of wonders all around you.

This box literally makes inaudible things audible. It also brings you an astonishing grasp of the invisible world: you can now understand with your physical senses as well as through your imagination and your learning. Once you get the hang of it – and I must confess that I am not on this level yet – you can identify the species from the sound you get bongoing from your bat detector. The frequency on which the bat is best heard is a good guide, and so is the pattern of the sounds you can hear: the three pipistrelle species make 'irregular smacks', the noctule 'alternating smacks and tocks', Daubenton's 'a series of clicks'; all this from the Bat Conservation Trust.

A basic bat detector will cost you a shade under a hundred quid – but have no fear. Many wildlife organisations offer evening bat-walks on which you can listen, and quite often have a go with a bat detector. I did one myself on a punt, going upriver from Cambridge to Granchester, with bats flying low over the river and shouting their torches of song. You can check out your local county wildlife trust, or look for a local bat group through the Bat Conservation Trust. The world is full of good and generous-hearted conservationists longing to show you the hidden life of bats: loving the bats and loving just as much the expression of astonished wonder that will light up your face when you find a bat for yourself.

There's one more magic trick you can perform with a bat detector. Turn the dial down, to the sounds that are only a little beyond our usual range, and you will quite often hear an odd grinding noise. You have found a cricket, or, more likely, crickets, because when you find one you tend to find several. These are males chirping competitively, seeking to attract females by means of the vigour of their song. They make the sound by rubbing body parts together and it's called stridulation. The cricket rubs one wing – equipped with a kind of rasp – against another with a straight edge.

So when the bats are quiet you can tune down your bat detector and discover that there are other forms of life just beyond the normal reach of your senses: other worlds going on all round your ankles, and you – before you learned this piece of magic – quite unaware.

7

The Bottomless Sit

'Yonder,' said Purun Bhagat, breasting the lower slopes of the Sewaliks, where the cacti stand up like seven-branched candlesticks – 'yonder I shall sit down and get knowledge.'

– *The Miracle of Purun Bhagat*, Rudyard Kipling

Bushbucks are small, handsome antelopes: a deep orange-brown broken up with white dots and dashes. They're about the size of large dogs. They're browsers (as opposed to grazers, eating the leaves of trees in preference to grasses) and you find them in thick bush, rather than out on the open plains. They're mostly solitary.

I was in an ebony glade in the Luangwa Valley in Zambia, sitting with my back to an ebony tree. I had been sitting there for an hour, scarcely moving. When I raised my binoculars, I did so in slow motion, as if I was doing tai chi, and lowered them with as much care.

It was a favourite sitting place. Every time I went there, events followed the same pattern. When I entered the glade, every bird in the place would take to the wing and vanish, shouting alarm calls, and the impalas – more antelopes, grazers, usually seen in gatherings of at least twenty – would bark their own alarm call and vanish. I had emptied the place in an instant. I had broken it.

The spell for mending was all in the sitting.

Bit by bit, life would reassemble itself all around me. Soon, birds would be foraging through the leaf litter at my feet; the impalas would drift back. It was as if I was slowly becoming invisible: my human nature drifting away into the landscape or being absorbed by the tree. And after a while my invisibility began to make other creatures visible. I was ceasing to matter.

And then came the bushbuck. A male: antlered, very dapper, and normally very wary. But – nibbling a bit from a bush here and a bit from a bush there – he eased his way closer and closer. And I just sat. Sometimes the most dramatic action of all is perfect stillness. We reached a point when the bushbuck was browsing less than five paces from my tree. His eyes passed over me once, perhaps twice, without seeing me: perhaps my spirit was absent. It helped that my outline was broken up by the trunk of the tree, and that I was wearing bush colours: khaki trousers, green shirt.

There was a long, perfect moment of impossible closeness – followed by the most colossal double-take in history. The bushbuck suddenly saw me for what I was and uttered a great, glade-filling bark of horror. He turned on his haunches and was gone. Everything else in the glade responded to the alarm, the birds fizzing up into the branches, the impalas vanishing. It was like being a leopard, suddenly spotted mid-stalk. I found myself laughing a little: what took you so long, all you brilliantly adapted, hyper-alert creatures of the bush? Where are all your super-senses? A stratagem as simple as sitting still had for a while confounded them.

And that's the spell. Sitting still. We humans are busy creatures, always getting on with something, always moving or talking or checking our phones. We have a terrible fear that if we stop for a moment we will miss something. The exact opposite is true.

Here's a little more from that Kipling tale quoted above, a story from *The Second Jungle Book*: 'Nearly all hermits and holy men who live apart from the big cities have the reputation of being able to work miracles with the wild things, but all the miracle lies in keeping still, in never making a hasty movement, and, for a long time, at least, in never looking directly at a visitor.'

The bottomless sit is best performed alone, which gets rid of the temptation to talk. But it can be done in the right company – the company of those who know how to keep still and silent. I remember sitting for a good two hours with a great lady called Margaret Grimwade, who was from the Suffolk Wildlife Trust, and my son Joseph. Sitting still for two hours is a hard thing to ask of a twelve-year-old, but he was up for the challenge. And eventually – eventually – our reward came along. It did so with striped noses and that glorious waddling gait: badgers, of course – what else? Margaret had placed us adroitly, with the breeze blowing from the badgers to us, so the badgers couldn't smell us. They were genuinely unaware of us: all we had to do was not move. And we didn't. So the badgers foraged and sniffled and snuffled and toddled and waddled, and we sat still, moving only our eyes – and neither Joseph nor I will ever forget it. No hide, no cover of any kind: our only disguise was our own stillness, and it worked triumphantly.

The spell of the bottomless sit works well when combined with the spell of the magic waterproof trousers, of course, but you don't always have them to hand. So here's a tip: put a plastic bag in your pocket. An ordinary supermarket carrier bag will do, but you can do better. In the pocket of my down waistcoat I keep a dry-bag, the sort of thing you use in various hearty outdoor activities when you want to keep precious stuff dry. But it's equally handy when the thing you want to keep dry is your bum. It's harder to sit for a long time when you have rising damp in your underclothes.

Introduce yourself to the bottomless sit by easy stages: after all, you're enjoying yourself, remember. At the start of every sit – especially when you're new to it – you will get fed up and restless pretty quickly: how long have I been sitting here ... Seems like hours. Whose idea was this? I thought there were supposed to be birds here? You may find your hand creeping towards your phone; you've just remembered something frightfully important and incredibly urgent.

But there's a moment when you go through a door. The urge to move decreases. Your eyes follow the birds, which are usually the most obvious bits of wildlife. The common birds hold your attention as they seldom do when you're walking: after all, there's not much else to look at. And – suddenly but subtly – you realise that time has changed gear on you. You're no longer waiting; you're just sitting, and you're not sure whether that's ten minutes you've been sitting or twenty, or maybe even more. You still hope that something fabulous will turn up, but you're far less bothered by this than you were when you first sat down.

I have had some remarkable sits. On the edge of a dyke near my home in Norfolk, a kingfisher once perched just behind me. I was able to keep him in sight by a very small movement of my head and by swivelling my left eye so far round it almost slid beneath my ear. I kept the bird in sight while it fidgeted, briefly preened, and then settled down for a good old stare at the water surface. Had I been double-jointed, he would have been within an arm's reach. Eventually he decided to change his hunting perch, as this one was doing him no good, so he flew past my head, wings not quite brushing my hair, the blue of his back so bright it might have damaged my retina, and assumed a new perch ten yards off.

On another occasion at the same spot I heard a splashing up the dyke. I was so used to the discipline of stillness that I even stilled the smile that was sneaking up the corners of my mouth as I anticipated what would happen next. And it did: a great bow-wave that reached both sides of the dyke, and, pushing it, a pointed furry head, hair combed back like a boy-racer and with the most expressive whiskers on the planet. An otter, no less.

These fabulous and privileged sightings are the point of this sort of sitting, and yet they're not the point at all. 'Teach us to care and not to care,' wrote T. S. Eliot in 'Ash Wednesday'. 'Teach us to sit still.'

Once you've had enough practice at sitting still, you learn to care and not to care. Of course you care about seeing the next otter, the next kingfisher, the next badger, but on the other hand, you don't really care if you don't see one at all. Being there is what counts.

Partly there's a meditation thing going on. Good sitting, sitting with a quiet mind, is a valuable thing that we seldom appreciate in the West. Maharishi Mahesh Yogi tried to get us to embrace his instant version of meditation; people seek it in yogic exercises in every church hall in the country – and yet inner quiet is a hard thing to find. But it's easier in a wild place.

Perhaps a quiet mind is easier to find when you're not seeking it for its own sake. If you seek the wild world, if you seek to rewild yourself by means of the bottomless sit, then you're probably seeking a kingfisher or some other special treat. The quiet mind is a by-product. Or is it?

Not every sit ends in a rarity. Many a sit will bring only the ordinary everyday wild things – but you find that you have moved a little closer to all wild things than you were before. You are becoming less an observer of the wild world than a living part of it. You're less of an outsider than you were before – and that's as good as seeing a kingfisher, maybe even better. It's not something you can look up in a field guide, or find on the website of your county wildlife trust. It's for you alone, and it's for doing, not for telling.

Save here. It's a secret and perhaps a slightly embarrassing one, but it's one that most people who love wildlife learn sooner or later. You tell yourself that you're out to hunt for this rarity or that, this charismatic species or another, but the truth of the matter is that it's wildness you're seeking. Not just the wildness you can see and hear and smell in all kinds of places, even in city parks. But also the wildness in you. The wildness that comes in the waiting. In the sitting.

8

How to Breathe Underwater

Harry clapped his hands around his throat, and felt two large slits just below his ears, flapping in the cold air . . . *he had gills.* Without pausing to think, he did the only thing that made sense – he flung himself forwards into the water.

– *Harry Potter and the Goblet of Fire*, J. K. Rowling

When you put your head underwater – in the right way, in the right place – it's not the access to air that's the problem, it's the tendency to gasp. My first attempt to do so as a grown-up lasted about a second and a half – that's when I realised I had gasped seawater and broken the spell. Fortunately, I was standing up in about three feet of water – the bath-warm water of the Red Sea – so I was able to retreat and try the incantation all over again. This time it worked. I submerged, I breathed, I swam, I entered a new country.

There is nothing in this book that feels quite as much like magic as this: to enter the right kind of water as a breathing and sighted creature and then to move away and meet those who really do have gills. All you need is a mask and a snorkel – and if you find yourself in a place where there's a bit of coral, you will be transported in a single instant into the land of the most incredible wonders you will ever see. It really is that easy.

Essentially a face mask is all you need, but it works better with a snorkel. A snorkel is just a tube, about as simple a bit of technology as you can come up with. You can wear fins – flippers – if you want to increase your mobility: worth doing because they make you feel so much more like a fish. But really, all you need is a clear watertight window between your eyes and the underwater kingdom. It will, at a stroke, cut out the surface glare, the ripples, the reflection and the distortion. All at once you can see plain in a place where you had never seen plain before.

You will have heard many times that rainforest is the richest and most biodiverse habitat on earth, and so it is. The problem is that most of the outrageously various creatures that live there are very hard to see. Most of the stuff you are looking for is up in the canopy: a hundred feet away at least, hidden behind a million leaves and branches. You can hear enigmatic calls – sweet and clear beyond description – but you've no idea what's making them. You may hear a crash and see a small shower of leaves: something's going on up there, but they're not letting you into the secret. Even when you're right in the middle of it, the biodiversity of the rainforest is something you have to take on trust: you're restricted to brief, thrilling hints and glimpses.

If you really want to get hold of the notion of the earth's richness, you must go to the sea. You must go to a coral reef and invoke the spell of the magic window – and then, in a single gasp-making instant, the riches of creation are laid out before you in dazzling colours and impossible shapes.

There's a strange sense of privilege that comes with this first miraculous vision. It's to do with the Age of the

Screen: the last century of human life. No one sees the skyline of New York for the first time any more – we've seen it in a thousand films. When you make your first trip to New York, it's like arriving in a place long familiar, as if you had entered a world of fiction, as if you were starring in a film of your own life.

We've all seen coral reefs a million times on television, or in the *Finding Nemo* films. When you see one for yourself, your first sentiment is naive shock: you mean this is *real*? You mean it's not just for television, and it's not just for the privileged? You mean it's something I can experience – I am experiencing – for myself? A new world of possibility opens up and you wonder if anything in the world is out of your reach.

This transition into the world beneath the waves is the most shocking through-the-wardrobe experience of them all. You shift in an instant of time from one form of existence to another. Even for the most timid of swimmers the experience is enriching and overwhelming. One look, just one gasping look, is enough to give you something you will remember all your life.

I'm a rotten swimmer myself, but I have seen the wonders beneath the surface of the sea on many occasions. I've never tried scuba-diving, intimidated by the kit and commitment, taking comfort in what a diver once told me: 'Snorkelling is much better. When you're snorkelling you're free. Just you and the water and the fish.'

Some people love the undersea life so much that they don't trouble with the stuff above the surface: it all seems unreal and unimportant. I know a family of four who go scuba-diving at every opportunity. They can't reliably tell a blackbird from a song thrush on Highbury Fields near their home, but an encounter with whale sharks enriched all their lives. It's all in the way these things take you.

I first learned the thrill of being able to see and breathe underwater when I was a boy. On holidays in Cornwall I would enter the rock pools, pull on my mask and trudge off in search of blennies and sea anemones: though it wasn't just about seeing them. I wasn't a fish-spotter, or an anemone-spotter: it was all about being among these alien creatures, being part of the same environment. Being a fish, inhabiting another world.

I know it's not the same as a tropical coral reef, but there are rewarding things to see off the coast of Britain. A good many underwater enthusiasts take photographs of what they find down there, and they soon get used to the usual response: 'You never took that here! That's from the Caribbean, isn't it?' But there are rich creatures and rich places to be found all over Britain: seahorses at Studland Bay, Dorset; sea squirts and seals at Prussia Cove, Cornwall; breeding spider crabs at Stackpole Quay in Wales; and on and on. There's plenty of good information a few clicks away.

The disadvantage of British waters is that they're seriously cold. And all this total immersion requires a certain degree of commitment, so perhaps you're beginning to think that this spell's not for you. But wait, there is a variation: a soft option, one in which you can remain more or less completely dry – and warm – while still seeing visions you once thought beyond your scope.

It's a bathyscope, sometimes called an aquascope. It's a gloriously simple bit of kit: not much more than a watertight window. You just lower it into the water, put your face in the appropriate place, and all at once you are seeing the same thing as a snorkeller: glare, reflection and distortion done away with.

They're easy enough to buy: just a couple of clicks and the willingness to spend around twenty-five quid. There are larger and slightly more expensive versions: you can get a bathyscope fitted with lights, so you can use it at night; that's a bit less than £100. There are collapsible versions, which are less of a fag to carry. You can use them from dry land or from a boat: boat people use them to check moorings; fishermen use them to check pots.

I remember using a bathyscope in one of those dismayingly swift rivers in the Lake District: wading across the bottom which was lined with perfectly round ankle-turning boulderettes. I was wearing waders, and the right one was gradually informing me of a rather serious leak. I was peering at the bottom in order to view the natural habitat of the freshwater pearl mussel, which is one of those intriguing creatures that radically alters the habitat it dwells in. They've become horribly rare; too many people have gone looking for freshwater pearls across the centuries. This was a rather thrilling project about reintroducing them.

So, as the water made a serious attempt to wash me out to the sea that lay a dozen miles distant, I was able to poke this device beneath the hurrying water and observe the way that the small boulders made their patterns and created minute opportunities in which, in the lee of a boulder, a mussel could anchor itself and so make a start on establishing a colony and enriching the entire river as it did so.

This was fascinating stuff, even though I was distracted by the state of my right foot and by the appearance of a dipper a few yards upstream. Mussels are great, but no birder can ignore a dipper.

Here, then, was the collision between two worlds. I and the dipper are both creatures of the air; below the surface there were fish and invertebrates. Astonishingly, the dipper – a perching bird just like a blackbird or a blue tit – is able to enter the water and forage between boulders for small scraps of life. It can fly through waterfalls, it can perch underwater, it can fly beneath the surface and then fly out into the air again as if it was all the same thing. And there was I in the middle of a dipper's stream, lowering my eyes – a little reluctantly – from the dipper and gazing down into the place where a dipper goes to work and makes his living – and I was able to see this different world with the greatest possible clarity. For a moment, I too was a dipper.

9

A Spell for Making Birds and Beasts Come Closer

'My sweet little sisters, birds of the sky,' Francis said, 'you are bound to heaven, to God, your Creator. In every beat of your wings and every note of your songs, praise him.'

– *The Little Flowers of St Francis*, Anon.

You probably know this spell already. If so, you'll know it also works for mammals: you can gaze into the eyes of a deer while the deer still thinks it a safe distance away. The spell is worked by means of a pair of binoculars, and if you think that's a little bit obvious, bear with me, because I shall shortly be telling you how to use them better: how to get more value, how to bring more birds still closer.

If you have binoculars, use them. If you don't, get some: a little advice on that also follows shortly.

The first thing about binoculars is not to be precious about them. Don't treat them like priceless crystal goblets: they are built to be reasonably robust, and to be used. Take them with you whenever you go for a walk; for that matter, whenever you go for a sit. You never know what might turn up; you never will if you haven't got your bins to bring the wild world a bit closer. You soon get used to the idea, and after a while it becomes unthinkable to walk without your bins: you would as soon go barefoot. Even if all you look at is seagulls, you will be rewarded.

Most binoculars come with a case and with straps and covers for the front and the back lenses. For a start, get rid of the case. All right, you can keep it for storing your bins in, but when you go out you don't need it at all. If you see something interesting, you want to look at it, you want to bring it closer. You don't want to start fumbling with a case: by the time you've got the bins out and up to your eyes the damn bird's flown off. Leave the case at home, or in the car or in your backpack. You need your bins naked and unashamed.

My wife says you should also keep a pair of bins in the car, especially if you live in the country. She has cursed herself for the lack on too many occasions, especially when there are hares about. Bringing the hares closer is never a bad idea.

Second, get rid of all those lens caps. The last thing you need to do before whipping your glasses up to your eyes to see if that passing pigeon was a falcon is to pop four separate caps off four separate lenses. Keep the lenses clean, of course you need to do that – the end of a scarf is useful here – but you don't need to shield them away from all the stresses of daily life. And, in particular, you don't need those lens caps when there's a chance of seeing something wonderful and you have a window of about two seconds to see it in.

There's one possible exception to the bin-your-lens-caps rule. That's the rain-guard. This is a bit of plastic that fits over the neck strap and hangs two or three inches above the back lenses. The idea is that it keeps the worst of the rain off, but falls away when you lift your bins to your eyes. I find them cumbersome myself and never use them – I keep tissues in my pockets and wipe the lenses instead. It's not entirely satisfactory, but it's OK.

Then there's the neck strap. Just about every pair of bins comes with a neck strap, and you can adjust it to your preferred length. Most people – other than committed birders – leave the strap as long as possible, so that the bins are swinging around just above the belt. This is both uncomfortable and useless. They get in the way, and you can give yourself some mildly painful blows when clearing a stile. But what's far worse is that their performance is compromised by the distance from your eyes. You have to lift them about two and a half feet before you can see through them.

So if you choose to wear your bins round your neck, keep the strap good and short. Some birders wear them so short the bins barely clear the birder's chin. Suit yourself, but remember that the idea of bins is to use them: to have them ready for immediate use.

I seldom wear my own bins round my neck. I find their weight uncomfortable. You can get a complicated shoulder harness to get round this problem, so that the weight of the bins is on your shoulders rather than your neck. I've never tried one: the idea seems to me confining and claustrophobic (like quite a lot of ideas, to be fair). I know people who use a harness and find it ideal. I carry my bins in my right hand, with the strap looped around my wrist for safety and from habit. When there's less chance of needing them in a hurry, I switch them to my left shoulder. Works for me.

People unused to binoculars lift them to their eyes and stick their elbows out as far as possible. That gives you the maximum opportunity for wobbling: and the stiller you can hold yourself, the better your view through the bins. Tuck your elbows into your sides, getting additional support from your torso, and you'll find you can hold them still for much longer. So there's a fairly practical bit of magic.

Some people feel self-conscious about using binoculars. Patrick Barkham, in his excellent book *The Butterfly Isles*, confesses to a terrible embarrassment about being seen in public looking through bins: fearful that everyone would instantly think he was a pervert. Be calm: remind yourself that there are more nature-lovers than perverts in this country.

Another reason for embarrassment about public binoculars is a fear that you are being pretentious: as if you are instantly making claims to an expertise that you don't possess. The first thing you have to understand is that no birder will take you for a crash-hot birder unless you are carrying top-of-the-range binoculars – and birders are as good at identifying each other as they are at identifying a drake smew from the opposite side of the reservoir. What's more, all crash-hot birders also carry a telescope and a tripod. (I never do that myself. Firstly, I find such gear too much hassle to carry about; secondly, I am not now and never have been a crash-hot birder.)

You are entitled to carry bins as a novice. And when you find yourself among other people with bins, apparently more expert than yourself, have no fear of asking what the hell they're looking at. It's a fact that showing another person a nice bit of nature is one of life's small pleasures. You show someone something marvellous – and a small bit of marvellousness is reflected back on you.

So OK, you are committed enough to want to upgrade your binoculars. Or perhaps, realising the lack in your life, you decide to treat yourself to a pair for the first time. What do you do? How much do you spend? Where do you go? You can get a pair for a few quid, you can get a pair for well over £1,000. Broadly speaking, the more money you spend, the better your bins.

But what's better? You'd think that the more magnification you can get the better. In thrillers, the characters are always looking at each through 'powerful' binoculars. Herr Flick, in the immortal television series *'Allo 'Allo*, used to remark: 'I am observing sroo my *powerful* Gestapo binoculars.'

But once your bins get beyond a certain level of powerfulness, they become useless. You can't hold them steady. You need a tripod to make them work properly. You need no more than seven, eight or ten mags (that's the first figure on the specifications of a pair of binoculars; I use a pair of 10x42s). Nothing bigger is any good.

So what do you get for your money? What's the bonus you get from expensive binoculars? The answer is light. Light and clarity. Optical devices suck light from the image they give you: the better – the more precise – the optics, the brighter the image. And there is also the three-dimensional qualities of the image. When a character in a film uses binoculars, we in the audience see a figure-of-eight-shaped image. But look through a proper pair of binoculars, and your image is circular. The clever old optics lay the image from your left eye over the image from your right, and that gives you proper stereoscopic vision. That's why your image appears fully three-dimensional. A good pair of binoculars is like being teleported across the intervening space. It's not like looking at something on a film: it's like being there. Good technology can be very like magic.

If you're looking for a new pair of bins to help you enjoy nature a little more, it's a good idea to go to a specialist binoculars shop. Not a photography shop that happens to sell a few bins as well: a proper specialist shop. You will be able to try a few pairs, and experience the difference between an OK pair, a good pair and a fabulous pair. You will also get proper advice.

I used to scoff, and think fancy optics was just kit-mania. Showing off. One-upmanship. Then I tried a pair of what were then top-of-the-range bins and gasped. And of course bought them more or less on the spot. So get the best you can afford, and use them.

That means the bins should be handy. If you might leave them at home because they are too heavy or too big, get a smaller pair. That might compromise you visually, but if you have a slightly inferior pair you always carry – it being no trouble – they're ten times better than the pair you left on the sideboard. My father uses a pair of miniature bins, and never takes the dog for a walk without putting them round his neck. As a result he sees a lot of birds. He wouldn't bother if his bins were as heavy as the ones I use. So it's about suiting your own lifestyle. Choose a pair you're going to use.

There's a drawback to even the best binoculars: you can't focus that close. On many occasions I have found myself walking backwards – away from the bird – in order to see it properly. And if you're interested in, say, butterflies, then you're going to spend a fair amount of time on the retreat.

But there is an answer: you can get close-focus binoculars. Their performance at long range is no better than adequate, but you can focus as close as eighteen inches, and that's a marvellous thing for the butterflies, especially when they have the decency to hold still.

These bins are another spell for taking you into a different world: the world of the small, the world of the invertebrates. Without needing to catch them and kill them, you can see them as close as if you held one in your hand. You can look at a parade of ants and wonder at their purpose. You can see small beetles on a flower head. You can see dragonflies in action. You can immerse yourself in the universe of the small. You need no special knowledge to do so, though you might wish to seek it afterwards.

So there I was on Alderney, one of the Channel Islands, a place rich in wildlife, not least because of the fine work of the Alderney Wildlife Trust. It's nearer to France than England, and that means that it routinely gets insects that are rarities further north.

And there was a hummingbird, and there was me watching it from a few feet away, through my magic close-focusing bins, and what I was seeing was a brilliant and glorious image. It was a hummingbird hawk moth, of course: an insect that flies from flower to flower and feeds from them at the hover, precisely as a hummingbird does. (It's a classic example of convergent evolution.) I had never found one for myself before: and there it was feeding away before me as if I was no threat whatsoever. Because of the magic bins, this lovely insect seemed to be performing specially for me.

So, in the warmer months I always carry two pairs of bins: one for the birds and one for the bees. One on my right hip, in a case looped onto my belt, and the other in my right hand. I have perpetual access to magic.

10

Travelling the Hidden Roads

Harry quickly took out his real wand and tapped the statue. Nothing happened. He looked back at the map. The tiniest speech bubble had appeared next to his figure. The word inside said *'Dissendium.'*

'Dissendium!' Harry whispered, tapping the stone witch again.

At once, the statue's hump opened wide enough to admit a fairly thin person . . .

– *Harry Potter and the Prisoner of Azkaban,*
J. K. Rowling

Come and walk with me across the floor of the Luangwa Valley in Zambia. There are roads everywhere, but no human made them. You can see them winding through the bush and across the areas of open plain: broad, beaten tracks that sometimes meander for no apparent reason, and sometimes proceed with the most obvious purpose from one point to another: well-used highways that take you from A to B with the greatest possible speed and convenience.

Who made those roads? Mostly elephants and hippos. There were rhinos here until the late 1970s: they too are great road-builders. And where they carve out their roads, others follow: it's always easier to follow a track than to force your way through thick bush.

Every evening the hippos leave the Luangwa River and set out across the valley to eat. They'll take all kinds of vegetation, but what they like best are the fruits of sausage trees: massive white things that give a sausage tree the look of an Italian delicatessen. When the fruits fall, they get devoured, and the seeds are spread along the hippo roads in the animals' dung. The roads often lead from one sausage tree to the next. Hippos tend to follow the same route every night: like many other creatures, they are creatures of habit. It follows that they make roads.

Look down at the river from the banks during the dry season and you will see on the sandy beach, between the banks and the distant river, a succession of paths. Hippos, elephants and every other large mammal that comes down to drink has a hand – or a paw – in making these paths. The steep bank itself contains a succession of gullies cut into the bank by hippos and elephants, and they can become very usable staircases.

Away from the river, the bush is a mass of pathways carved out of the landscape by the huge population of large mammals that live there. They are created by habit and by need, and they are maintained by the same two things. It's what mammals do.

It follows, then, that the mammals that live in our own country do the same thing. It's harder to pick out the wild roads over here, because there is so much human interference, and because most of the mammals are comparatively small and so they make smaller and less obvious roads. But once you start to look, you find yourself picking out the hidden pathways: and you are beginning to find an understanding of the place you live. The parallel existence of mammals other than ourselves or our domestic beasts is revealed by subtle changes in the vegetation – but not so subtle that they're beyond your scope.

Try walking along a country lane in summer, when the

vegetation is high. Keep half an eye on the growth on the side of the road. You might find a line crossing that vegetation, a line that looks a little darker than the stuff on either side. Look a little closer. You may see that the plants have been shouldered aside. You may notice that the floor of this path, narrow though it is, has no tall plants in it.

So here's the clincher. Look on the opposite side of the road. Is there another dark line, another hint of shoved-aside vegetation? If so, you've almost certainly found something. It's not a path in any human understanding of the term, but it's a highway for those that use it: one that takes them from one side of the human-made road to the other. It's a crossroad: an intersection between the wild world and the tame one.

The path may well have a roof over it. That's often what stops us recognising it as a path – that and the habit of not noticing. We humans stand five and six feet tall or more: our brains can't cope with the notion of a road that doesn't accommodate us. Make a small mental shift and you will be able to pick out the handiwork – the pawiwork – of our fellow mammals where you never did before. A small roofed path, more or less in places a tunnel, perhaps a foot wide and two and a half feet high – this will often be the work of small deer: the introduced Reeves's muntjac and Chinese water deer, or the native roe deer.

Keep in mind the principle that, like you, most mammals – most animals – will take the easy option if there's one available. An established roadway offers a convenience that's hard to turn down, and so it gets used again, making it still more useful. And obvious. It follows that human-made paths and roads are often used by our fellow mammals. I have seen lions using the dirt tracks made for the Toyota Land Cruisers that carry the Luangwa Valley's visitors from wonder to wonder. The only time I ever saw a jaguar was on a vehicle track. At my place in Norfolk we maintain a path so we can walk around the marsh and enjoy it. It's used routinely by the mammals we share the place with. They leave signs: turds and footprints.

As you take a walk, seek to be aware of those places where the plants don't quite look as if they grew that way. Of course, you have to take into account the fact that you're not in the bush and that other humans and their dogs may have been there. In some places there are sheep, leaving their own signs when they're not standing in the middle of the road looking daft. It's harder to find wild in this country than in Zambia.

But, as you acquire the right eyes, you will find yourself doing so: a discolouration on the forest floor might be a path, so have a look at it. Check it out: it might have a few conical pits at irregular intervals, dug at an angle, about the length of your first two fingers. Does it feel like a good round shape, a little flat at the bottom? If so, smile: you may have found the snuffle-pit of a badger. A badger's response to anything interesting is to make a hole: the invertebrates they dig up are the most important part of their diet. Now tell me: did you find a hair in that pit? If so, take it out and then rub it between your finger and thumb. Does the hair seem to have corners? As if it was a square in cross-section? If so, then you can congratulate yourself: you've definitely found your badger.

Or you're strolling along a river, a stream, a dyke. You notice that there's an animal path leading off the path you are treading yourself and heading in the direction of the river. Follow it a little way, if you can. Perhaps where it ends you find a small area – say a couple of feet square – of flattened grass. And the vegetation by the edge of the watercourse had been pushed down so that it hangs over the lip of the bank. Another moment of excitement: you have found an otter. This is a place where an otter will occasionally rest up – and where it will also slither back into the water. It's an otter slide.

It's not just these – comparatively – large mammals that make roads. If you find yourself walking through rank, tangled grassland, it's worth taking a look down at ankle height. It's not too hard to find a small entrance into the sawn-off jungle. Beneath it there will be a network of tunnels: these are the highways of the short-tailed field vole. They make their living by tunnelling *above* the surface of the earth, feeding on vegetation in the security of the grass-roofed tunnel system they know like the back of their paws. They will mark their trails with urine, which is both territorial advertisement and navigational aid: they can find their way through their tunnel systems by smell.

It's not as secure a system as they would like. Barn owls can crack it by means of the silence of their flight: they can hear the voles moving below them as they make an entirely noiseless progress above the patch of grass. Their ears are set asymmetrically, so they can get a cross-bearing on the sound and pounce with confidence on a beast they can't see. Kestrels can also crack the tunnel system, but they use a completely different method. They can see the ultraviolet fluorescence of the urine trails the voles leave and can use them to pinpoint the whereabouts of the vole.

But the tunnel system works well enough to allow the voles to prosper. And as you train your eyes and your mind, as you become that little bit wilder in your heart, so will you become aware of their roads, as if you were a kestrel or a barn owl. So you will become a little more intimate with the creatures you share your countryside with.

11

How to Look Beyond the Edge of the Earth

There, as in a mirror, he could see, at certain times, what was going on in the streets of cities far further south than Tashbaan, or what ships were putting into Redhaven in the remote Seven Isles, or what robbers or wild beasts stirred in the great Western forests between Lantern Waste and Telmar.

– *The Horse and His Boy,* C. S. Lewis

Some great magicians possess the skill of extending their visual awareness beyond the confines of the space they occupy. In his magic pool, the Hermit of the Southern March witnessed the distant battle at the Castle of Anvard, in *The Horse and His Boy*. It's as if they are able to hurl their very souls from them, without losing contact with their own bodies, their own senses. Modesty Blaise, heroine of the great thriller series and adept at yogic meditation, claims that she once extended her visual awareness beyond the room in which she was meditating.

This is the essence of the next magic spell I offer you. It is an adaptation of the bottomless sit, usually involves binoculars, and there are occasions when a pair of magic waterproof trousers is a great help.

We have already examined the life of the sea, and the startling new understanding of its possibilities that comes with even a short voyage. Here is a way of doing the same sort of thing but without going near a boat. However, it helps if you have done the boat thing before you try it. That will change your expectations and allow you to see a good deal more.

Sea-watching is a specialised activity. Not all birders go

in for it. Those who do tend to be pretty damn good. The job involves staring through a telescope for impossible lengths of time. It's also about interpreting shockingly meagre visual clues to achieve genuine identification. A further art is to pick a good day for it: best by far when fierce on-shore winds blow the birds of open ocean towards land. Experienced sea-watchers talk with nostalgia about the great storms during which they have sat on the edge of the sea and gazed without pause for hour after hour. The whole business is about going out to sea while staying on dry land.

There are plenty of profitable and amusing ways to perform this piece of magic without much expertise. And you certainly don't need a scope. You can do all right with naked eyes, but you'll do much better with bins.

So find your observation point. Logic will tell you that the more you stick out into the sea, the close you'll be to the seabirds. It follows that a headland or the end of a pier are good places. Height is good: so a nice clifftop, should one be available, is just about perfect.

Take your position, make yourself comfortable – and then throw your soul out to sea. By that, I mean focus your mind, your eyes and your bins on an area half a mile to a mile away. A place few people on land concentrate on. And then . . . see what happens. It may not be very much, almost certainly not at first, but you will find the process of sitting and staring out to sea, in the manner of the French lieutenant's woman, oddly pleasing. Looking at water, like looking at fire, is the thinking person's television.

The easiest seabird – not counting the shoregulls that we're all familiar with – are gannets. They might have been designed for the pleasure of sea-watchers. They stand out from the sea with startling whiteness. Even on a day when the wind flicks white tops onto every wave, the gannets are whiter: detergent-advertisement white, a white that makes my whites look grey, as the old slogan had it. Also, they will hold that perfect cruciform silhouette. That's very eye-catching, and the flight is distinctively glidy. You can pick them out with naked eyes even at a distance: raise your bins and you'll see the black tips to the wings. Sometimes they will travel in parties of half a dozen or more: almost always low to the water in a neat V-formation, or in line astern.

Nearer inshore, you'll probably make out cormorants and shags: black birds, usually on their own, sitting low on the water, and as soon as you turn your bins on them, they've dived. They're always worth watching, even at a distance, because of the pleasure of telling one species from the other. When a cormorant dives – both species dive from the surface, in the manner of a duck, not from a height, in the manner of gannet or a tern – it slithers neatly into the water, like an otter. But when a shag dives, it does so with a gamesome little hop, bouncing a fraction clear of the water before going down. It's a very cheering thing to see, and there are opportunities for a thousand bad jokes if you choose to tell others of your sighting.

Look out also for rafts of ducks. You can normally distinguish them easily enough from gulls, because the white of the gulls shows up even at a distance and even when they're not much more than silhouettes. Besides, the gulls ride in the water with a high stern: sea ducks keep their back ends down low.

We're used to ducks as freshwater creatures, but some are versatile – you can often see mallards in coastal inlets, especially in winter. Some species specialise in the sea. Look for a nice raft of ducks and try to stay with it. Here you will find one of the great snags of sea-watching: the waves do rather tend to go up and down. First you can see your ducks plain as anything; an instant later they've gone, and you can't find them again. Then there's a shift, either in their position or more likely in the rhythm of the waves, and they're visible again. Most frustrating if you're aiming for a serious ID.

But if you're just there for the pleasure of throwing your soul out to sea, that's a small matter. If they're dark – if they seem more or less black – there's a good chance they're common scoters. If you're with a non-birder, pronounce that name with confidence. If you're picking out the occasional splash of white, then you're more likely to have some eiders with some males among them. Remember, however, that most of the joy is in the looking, and in the looking *for*. If you can't stand not being sure about ID, then maybe you'd better avoid this spell. But I would sooner you learned the pleasures of nature, and the pleasure of immersing yourself in it and embracing its puzzles and its deeper mysteries. Good ID is great, but it's not everything.

It's also pleasing if you find a shearwater. Unless you're a crash-hot Premier League birder with a top-of-the-range scope, you probably won't be able to tell what species of shearwater it is, but never mind: there's a great thrill in picking out one of these lovely birds as you sit dry-bummed on the land. The name is wonderfully apt: you can see them, even at extreme range, as they turn and bank, standing on one sharp-bladed wing, its tip almost touching the sea. Its wonderful elegance at so great a distance is a secret that you – and you alone – have managed to crack.

That too is part of the pleasure. A little sea-watching makes you an initiate. The people who walk the cliff path behind you and give you a polite greeting as they pass – they have no idea that there's a shearwater or a sea duck or a gannet out there. They're walking by the sea but their souls are still tied to the land. They look out at the sea, but they see only what's close to the shore. They're coastal people, but you're oceanic. You're pelagic.

I remember one summer when we took a beach hut in Southwold, which is the beach-hut capital of England. It was called Sandpiper. I established a very pleasant rhythm. For a while I would descend to the beach and play Back Goes the Sea with my younger son, as fine a game as any that exists. Then I would retreat to Sandpiper, open a beer and stare out to sea. After all, you never know what will turn up – and even if it doesn't, you've spent some quality time with the ocean. I saw many fine things in the few days of Sandpiper, but the one I remember best is the Arctic skua. There it was, black as your hat with white flashes on the wings and a discreet black streamer in the tail. And, of the thousands of people all along the beach, only I knew. I had shared something with that bird, even if it was only my own awareness. And I was enriched by it.

Staring out to sea is great, but staring out to sea with an expectation of seeing something changes the game. It means you are much more likely to see something. I was at a clifftop pub in Cornwall, near Tintagel; the table held my pint and my bins, and the pint of my old friend Ralph. It didn't trouble him that I spent a good deal of our conversation with my eyes focused somewhere between half a mile and a mile away. He has known me for fifty-odd years, after all.

And then, at once:

'Dolphin!'

I remember shouting the word at the absolute top of my voice. It was obvious, essential, that everyone within my earshot should turn and, like me, look to sea. And there they were, three of them, common dolphins, seriously large animals, cruising across the ocean in that glorious series of arcing sigmoid undulations that dolphins use for speedy A-to-B travelling.

I remember one other thing. A second after they had gone, a woman who had seen the dolphins because of my shout turned and looked up at me from the road that ran below the pub. Catching my eye, she gave me a smile that might have slayed me. It wasn't a sexy thing, but it was a moment of real beauty, her smile lighting up her face and the day, much as the dolphins had been lighting up the day a few moments earlier. And I was struck, as if by lightning, with the joy of showing someone a bit of nature: the sharing, the showing, the gratitude, my gratitude for her gratitude, the lovely illusion that the wonder of the dolphins was something to do with the wonder of me. Nothing of the kind, of course, but in the showing there is an act of profound pleasure: sharing with her, sharing with the dolphins themselves.

And that's why I'm writing this book.

12

How to Penetrate the Darkness

'But what manner of use would it be ploughing through that blackness?' asked Drinian.

'Use?' replied Reepicheep. 'Use, Captain? If by use you mean filling our bellies or our purses, I confess it will be no use at all. So far as I know we did not set sail to look for things useful but to seek honour and adventures. And here is as great an adventure as I ever heard of.'

– *The Voyage of the Dawn Treader*, C. S. Lewis

I had never seen such a thing before, and I was filled with joy and wonder. I looked it up in the book, identifying it without trouble – it was there on the cover, no less. It was then that a single word overwhelmed me. It was the last word I expected to find, and it was a word that told me I had to write this book.

The word was *common*.

How could this amazing thing be common? If it was common, how come I hadn't seen such a thing many times over?

It was a moth. I expect we could all make that diagnosis, even if it was big enough to count as an honorary bird. We expect moths to come in fifty shades of brown: this one was as bright and beautiful as any butterfly. It was mostly green and pink, but in very subtle, even tasteful shades. It stood on the tip of my finger, its wings covering about the same area as the palm of my hand. It was so big I could feel the firm grip it took on my fingers. Confession: I'm not entirely at my ease with large insects, and have a sort of mid-range arachnophobia, so I felt a certain degree of discomfort at the legs and the grip, but the beauty of the creature forced me to get over such feelings. It was, after all, a moment of perfect privilege.

It was an elephant hawk moth: so called because the caterpillars – sizeable things, as you would expect – look a bit like the trunks of elephants. These moths are likely to turn up anywhere you can find rosebay willowherb, which is the caterpillar's food-plant. And that's practically everywhere.

The reason I had never seen one before is that elephant hawk moths fly by night. (There are plenty of moths that fly by day, most notably the gorgeous – and common – six-spot burnet, which has six red spots on each green wing, and is easily seen on flowery grasslands and clifftops.)

Night is another world. We don't live there. But bats do, as we have already seen – or heard. And so do many species of moth. We can find bats with a bat detector; we can find moths with a moth-trap. There are home-made Heath Robinson devices you can set up to attract moths: the easiest is to hang a white bedsheet from a washing line and shine a bright light at it: you can get a suitable torch for a few quid and prop it up in front of the sheet.

Or you can go the whole hog and get a moth-trap. I was given one for a recent birthday, and right from the start it was obvious that this was not so much a piece of kit as a magic spell. You can get one for £100, and they go up to around £140 or a little more. That's quite a commitment, of course (though, having spent so much, you tend to use it just to get your money's worth).

But you don't have to get one of your own. Not at first, anyway. You can experience at least some of the joyous discovery these things bring you by going along to a moth-trapping event run by your local wildlife organisation. Most county wildlife trusts will put on one or two such events in the warmer months, and usually they're designed for beginners. They're not for addicts and adepts: they're trying to get you hooked. Or, at least, to penetrate the night a little more fully than you did before.

A moth-trap is not a sophisticated device. Basically, it's a bright light: a mercury vapour light, to be technical. This sends out a light of peculiar brightness, with a somewhat eerie quality. It's as if a UFO had landed in your garden to Make Contact. (So it's as well to ask permission of your neighbours before turning it loose.)

The light is placed over a box, one that's easy to slide into but quite tricky to climb or fly out of. (They can manage it all right, though, so the earlier you are at your trap after daybreak, the more moths you'll find.) Inside the box you place a few egg boxes: the moths can creep into individual cells and shut down, with no danger of harming themselves or each other.

And what you do is get up, go down to your trap, switch off the light, and marvel. That's the easy bit. The harder thing is to identify them.

So, let me come clean. I'm hopeless at it. I literally don't know where to start: which page to look on, even which part of the book to start looking. I'm like a birdwatcher who has no idea that there's a clear difference between swans and robins; I have to thumb through the whole damn book to even get a clue. Expertise, then, is not essential for the enjoyment of moths. That's true of every aspect of wildlifing: the aim is wonder and delight, rather than a serious addition to the world's scientific knowledge. However, the acquisition of personal knowledge – however meagre – is always a thrilling thing, no matter what age you reached at your last birthday. Learning even one new name, one new moth, is a small and thrilling adventure.

If you have a smartphone with a camera, you have a useful tool for moths. Most of them come out of the trap in a pretty torpid state, so you can take a decent picture and look them up at leisure. (You can get transparent plastic containers for moths, so you can examine them at greater leisure. Specialist suppliers sell them and they're cheap.) As you empty your trap, you must make sure you place each moth in good cover. It's not a hard job and you don't want to make a bird table of all these beasts you have summoned to your trap.

Moth-trapping has been described to me as Christmas Day: you approach your moth-trap with no idea what you've got for Christmas. It might be a pair of socks; it might be the Kohinoor diamond. Or both, of course.

For the beginner – and no doubt also for the true adept – even the socks are wonderful. You can look at these creatures not only close up, but still, often enough on your own hand. And as you look, you can see the exquisite nature of the ordinary moths. They're all as gorgeous as butterflies, but rather subtler. Identification is a lot harder than butterflies – as I already said, there are only fifty-nine species to worry about in this country. There are, however, around 2,500 moths in the UK. True, most of these are micro-moths – the very small ones – and they're very much the domain of the specialist. The rest, the bigger ones, the macro-moths – that's a rough-and-ready distinction – number around eight hundred in this country, which is still an awful lot.

So it's all about trying to identify the easy ones, having a stab at the slightly less easy ones, and accepting that there'll be a lot you're going to miss at first – and perhaps for a long time or even for ever. But no matter, it's a great way to appreciate the notion of biodiversity without going to a rainforest.

Look at your trap, or at the trap of your local wildlifers. So many individuals: so many different *kinds* of individuals, so many *species*. And that's not just a conundrum for humans trying to put a name to them – it's also a thrilling revelation of the way that life works, which is by making lots and lots of different kinds of things. Different species. And every single species nestling there among the egg boxes represents a different way of life. That's a head-spinning concept. What we once covered with a single word – moth – represents 2,500 different ways of earning a living and running a life in this country alone. In the entire world, there are around 160,000 species of moths, along with 17,500 species of butterflies. And every single one of those species goes about the problem of living in a slightly different way. Not only different, but equally effective. Each one has a strategy that has allowed it to survive from one generation to the next.

We had an elephant in our garden, as I have said. We also had a tiger. Two or three of them, actually. This is a moth as gaudy as any butterfly, but in a slightly delicate way. The forewings, which slide back over the rear wings at rest in the classic moth fashion, are black and white in a sort of op-art swirl-pattern. But when it shifts these wings forward, it reveals hind-wings of startling orange with blue-black spots. This is a moth called the garden tiger, and though you may never have seen one in your life before, they are also classified as common. The caterpillars feed on nettles, docks and many garden plants: generalists often do well around humans. One moment with a garden tiger justifies all the expense and trouble of running a moth-trap.

There was also a fractionally more subtle one that I warmed to. It was in a gorgeous shade of minty green, and it's called a common – that word again – emerald.

There are elephants at the bottom of the garden. Also tigers. Also emeralds. There are miracles at the bottom of every garden: so many of them just below the threshold of the average human's awareness. But you can change that. You can cease to be an average human.

13

Time Travel

'It's called a Time-Turner,' Hermione whispered, 'and I got it from Professor McGonagall on our first day back. I've been using it all year to get to all my lessons.'

– *Harry Potter and the Prisoner of Azkaban*,
J. K. Rowling

You don't need special equipment to build your own time machine. You have a TARDIS in your pocket, and it will take you through Time And Relative Dimension In Space. It's an alarm clock, or the wake-up function on your phone, or whatever it is you use when you need to get up earlier than your body would like.

We default to our origins with very little prompting – but that prompting has to come from somewhere. Back in the days when we hunted and gathered, dawn was our best time.

When I go back to the savannah, I get up at dawn, not as if but because it's the most natural thing in the world. Daylight is a summons to action, to life. It's no hardship: there's a sense of privilege in stepping from your hut into the soft light of an African day before the sun has shown itself. It's the best time because it's cool. That's nice for you, but it's also nice for all your fellow species of animal that live out here in the bush. The middle of the day is too hot: the elephants and the antelopes and the lions doze under trees and bushes while you doze in your hut. You'll be up for more adventures when the sun is three parts of the way back down again, and so will the elephants.

If you live without artificial light, dawn is the most important time of day. It's safer, infinitely safer than the night you've just lived through, and so you can start reminding your neighbours that you're about and that this happens to be where you live. This occurs in various ways on every day of the year – and yet we humans snore our way through it, cut off from the natural rhythms of the world, the natural rhythms by which our ancestors lived.

But we can return to these times. We can renew ourselves by greeting the dawn. And though such an adventure will have its rewards at any time of the year, I suggest 1 May. That's if you can stand the din.

The latest you can leave it is 3.45, and that's only if you have a wild place on your doorstep. You need to be out in the right place before 4am. Oh, and a word of advice: dress for a cold winter's day. Those magic waterproof trousers are not a bad idea either, no matter what the weather. There'll be plenty of dew.

You'll be lucky to get a complete absence of man-made sounds, but you'll get a quietness that's unknown at almost any other time, and that in itself is a powerful experience. We are so used to shutting out extraneous noise that, when there isn't any, our brains need to make an adjustment. There's something faintly alarming, faintly thrilling, in this comparative silence.

What you do next is up to you. You can walk, or you

can sit. Either is good: it's a question of taste and temperament – and, also, of the place you happen to have chosen. If you can find a single spot that's full of promise, you can sit tight. Otherwise you can walk, but slowly and with frequent stops. The best of the experience comes long before the light is bright enough to see easily.

The dawn chorus is perhaps the single biggest wildlife miracle that we have in Britain – and it's open to us all, no matter who we are or where we happen to live. The birds do it throughout spring, but they reach their peak round about May Day. The resident birds will still be hard at it and the migrants would have by now set up territories and will be announcing them and defending them by means of their voices. By means of song.

Let me stress right away that you don't need to be an expert to appreciate this. I will be getting on to the spell of birdsong later in this book, and it will allow you the great adventure of our forgotten sense, the sense of hearing. Your local wildlife organisations will certainly offer dawn-chorus walks, which are companionable and come with experts who have ears like fennec foxes, though miraculously miniaturised. There will often be breakfast or at least coffee.

But you can also do it for yourself. You can greet the dawn in a silence that is stripped of politeness and friendly put-yourself-at-ease jokes. You can let yourself go. You can let the dawn itself take over ... and one by one the chorus swells, till it's a mighty noise, as Mike Heron of the Incredible String Band sang.

When the world is charcoal-grey and the trees are silhouettes and you can't believe it could ever be this cold in this country in May – so the singers wake one by one and sing their hearts out in the biggest and most important song of the day. I'm still here! I'm still alive, and I want my hen nearby and all those cocks within hearing range to know that I'm still alive and I'm still in my prime. And all around and on every side the answer comes back: well, so am I!

Each individual paired-up male must sing or perish, and each of the many species must send out the same message but in different music. With practice you can easily tell which species is singing, but never mind that for now. Just lose yourself in the music, as those birds have surely lost themselves. They're not thinking about sex and territory and becoming an ancestor: they're here in the present tense and they sing, and what they sing is the song of life, the music of the spheres, the sound that drives the earth around the sun.

So if you can't tell a crow from a blackbird by means of your ear, don't worry: just be out there when the music starts to play. Listen to the layers. Think of the number of individuals, the number of species, the number of singers, the number of songs.

As the light slowly advances and colour starts to be a

factor in your life again, keep looking. The sharp end of the day can produce remarkable and unexpected sights. Some of the strictly nocturnal mammals are finishing their work for the night; other mammals are about and even visible because they have learned to like this time. Not only is it the time of new beginnings, but there are very few humans around. Between first light and first humans there is a gap of two or three hours at this time of year, and it's a good plan to make the most of them. Dawn vigils have brought me some remarkable encounters. I remember open heathland in Suffolk that seemed to have been turned for a brief while into the savannahs as herds of red deer claimed it for their own. I was hidden in a vehicle at the time, which made it easier, and I knew that by the time the roads were humming again the deer would have returned into cover.

This was their time, and it was like being in another country. Indeed, it more or less *was* another country. An Israeli once explained to me that that the Israelis didn't need to build parks and other places of peace and quiet into their cities in space: instead, they built them in time. The Sabbath. During the Sabbath, even crazy old Tel Aviv is a place of gentle calm.

You can do the same thing for yourself. You can find a place of peace and quiet by means of your alarm clock: that place will be full of the wonders of the wild world.

The night itself is also such a place. If you get out into a wild place in the night, you can penetrate the darkness by means of your ears. In many places in southern and eastern England you might hear the barking roar of Reeves's muntjac. At many times of year – but most often in spring and autumn – you will hear owls, most often tawny owls. Everyone is familiar with the long, wavering hoot of a tawny, because you hear it in every graveyard scene that was ever filmed. But tawnies make another sound: a sharp, two-syllable call with a strong stress on the second: kee-*vitt*! Tawnies don't go to-whit-to-woo; they go to-whit *and* they go to-woo. The hoot is normally a declaration of ownership: this is my territory. The second call is more a contact call: I'm here, where are you? But they can be mixed up and they vary very much in intensity. Owls are creatures of the night and when they need to attract each other's attention they yell.

I remember a fabulous nocturnal concert I once attended – *eine kleine nachtmusik* – again on the heaths of Suffolk. It had five principal vocalists, taking turns, sometimes calling in twos and threes and occasionally all together: an enthralling performance and I sat and listened for an hour. The main performers were nightjar, with its mad radiophonic song, and the incredible complexities of nightingale song. This was mixed with stone curlew and the distant bass thrum of a bittern from the reedbeds beyond the heath, and . . . what the hell was that? A sort of gargling, clicking sound? Natterjack toad.

And all this about twenty yards from the road, with occasional passing vehicles full of people who had no idea that this wonderful stuff was going on outside the sacred spaces of their own vehicles. I wanted to flag them down and force them to listen: this is one of the most wonderful things you'll ever hear, if you but knew it!

I thought on the whole that I'd better write this book instead.

The spell of the dawn is the best, though. The notion of a familiar place going all sinister on you is a familiar device in films: the sudden draught, the fluttering curtain, the shadow on the wall . . . well, here's the reverse: a familiar spot that becomes, by means of a brief journey through time, a place lit up by the flames of eternity – oh brave old world that has such singers in it!

14

Magic Words

The Queen let go of his hand and raised her arm. She drew herself up to her full height and stood rigid. Then she said something which they couldn't understand (but it sounded horrid) and made an action as if she were throwing something towards the doors. And those high and heavy doors trembled for a second as if they were made of silk and then crumbled away till there was nothing left of them but a heap of dust on the threshold.

– *The Magician's Nephew,* C. S. Lewis

For everything and everyone in the world there is a magic word. It changes your relationship for ever, increasing intimacy and understanding at a single utterance. It works for people and it works for domestic animals and it works for wildlife. There is a magic word for each individual and for each species, and once you learn it you will never look at the person or the creature in the same way ever again.

The magic word is a name.

There is all the difference in the world between that woman next door and Mrs Ada Cramp. There is a world of difference between that bloke in accounts and Reg. If you regularly go to the same pub, the landlord will make a point of learning your name, and will be glad if you start using his. I used to go to a pub with my then neighbour, Eddy. The resplendent Gladys Trim, prima dominatrix of the Licensed Victuallers Association – I'm not sure if that was her exact title – always remembered my name, but never Eddy's. 'Good evening, Mr Barnes!' she would greet me warmly, adding kindly to Eddy: 'Hello, dearie.'

The magic of names also operates with wildlife. Learn the name, and a new relationship begins instantly. I have already touched on this, with the magic of the buddleia bush and the moth-trap. Suddenly one word – butterfly – is no longer enough. An early spring day is made still brighter when you see not a hurrying butterfly, but a gorgeous – resplendent – orange-tip. That's not a hard name to remember: the white wings of this butterfly have bright orange tips (at least, those of the males do, but let's not get too complicated).

In the same way, the words 'moth' and 'bat' rapidly become inadequate when it comes to striking up a relationship with the flying creatures of the night. It's nice to see a big pink moth, but it's more meaningful to see an elephant hawk moth. Even if your research takes you no further, the fact that you know the name has already made a difference. It means you have picked it out from eight hundred species of large British moths, and it means that you will probably recognise it again. You have made a place in your brain for the elephant hawk moth: you have stored an image and you have secured it with a name. In other words, the moth has become part of you.

Now let's turn to a field guide for birds – that is to say, a book that tells you how to tell one species of bird from another. You probably have one: if not, I suggest you get one at once. (Hint: get one restricted to British birds. Save Europe – and North Africa – for later.) You may not use your field guide all that often: occasionally looking up a bird that comes to the feeders, perhaps. And when you do look, you might find that the book is full of birds you have never seen. Their names are mostly unfamiliar and even a little irritating. The book is rather off-putting.

So let's deal with that. Near the beginning of every proper field guide you will find ducks – rather a lot of them. We'll start with them, not least because they have some funky names: mallard, wigeon, pochard, gadwall, teal, shoveler, goldeneye, goosander, smew ...

I suggest you peruse these names, and the pictures that go with them. You can do the same with the other birds in the book. Do so in an idle, random way, thumbing through, enjoying the strangeness of the names and the obscurity of the birds named. Leave the book in a place where you might routinely give it the odd five minutes – by your favourite chair, bedside table, in the loo – brief moments of painless study and a process that makes the strange names familiar.

I did this as a child. I read *The Observer's Book of Birds* like a favourite novel, till I more or less knew it by heart ... and this was years before I learned how to look for real birds in real places. I knew the magic of the name falcon, and I knew that falcons came in species like peregrine – that one word was in itself an adventure, all the more so when I learned that the name means wanderer or pilgrim – and also hobby, kestrel, merlin.

The next step is to put a face to the name. That means seeing the bird for real, moving about and living its life, rather than posing for a picture, almost always facing hard left. I suggest you take a nice walk around a lake. You'll be able to find a good one nearby by checking out your local county wildlife site and the RSPB, and they'll even give you a clue of what birds you're likely to find. Winter is the best time for ducks: instead of being spread out and breeding, they're all together in flocks, and the British birds are not alone: ducks come to enjoy our balmy winters, fleeing from Scandinavia, Eastern Europe and Siberia, some birds coming down from their breeding grounds inside the Arctic circle.

Ducks are both colourful and obliging. They sit still on the water, rather than flitting through the branches. Concentrate on the males: they are brighter and much easier to identify than females and immatures. Don't worry about the birds you can't put a name to. There'll be plenty of these, usually dismayingly far out on the water: well, just enjoy the numbers and the enigmatic patterns and colours. There'll usually be a good few reasonably close in though, allowing you to have a good old stare and eventually to find a name. And so change your relationship for ever.

It's got a green head, it's a mallard – but you probably knew that. You'd think from the illustration in the book that a teal would be unmistakable from miles off, but often enough, you need a good light to see that lovely head pattern. Is that grey duck a female of something? But it has a black stern. It's a gadwall. A group of ducks with orange heads: if they've got black breasts, they're pochards; if they're pale, they're probably wigeons.

It really can be this easy. I went to Abberton Reservoir in Essex on a January morning and saw twelve species of ducks before lunch: that's not counting the other birds sitting on the water: geese, grebes, coots, gulls and cormorants. That pale duck with an enormous beak: that's a shoveler. That almost pure white one with a sleek, elegant look to it: a goosander.

Finally – and this one's easy to tell, because there's usually a dozen people staring at it through cameras and telescopes and binoculars – there's a smew: a bird that crops up in winter, normally during a cold snap and mostly in the southeast of England. It's a dashing little thing that floats in a manner not totally unlike a celluloid duck, almost completely white, but with stylish black markings, especially over the eyes. Who was that masked duck? The lone smew swims again . . .

Now you have the names and the ducks themselves safe in the treasure-house of your mind. You've had the pleasure of seeing them, and with a bit of luck you'll know them next time. And even if you don't, you'll still be much quicker finding them in the book.

But don't stop there. Carry on thumbing through that field guide, revelling in images and names, and you'll notice – but maybe your eyes glazed over – that there are other names: secret names, aliases, assumed names, code names. These are scientific names, but they're not only for scientists.

'She's called Penny,' a man said of his spaniel, in a rather showing-off sort of way. 'After my favourite duck.' I was supposed to ask for an explanation, but I had no need.

'Good job your favourite duck isn't mallard, then,' I replied.

This was a rather fine piece of one-upmanship, and I suspect that the great Stephen Potter – author of those seminal works *Gamesmanship* and *One-Upmanship* – would have approved. The wigeon is a lovely duck, and if you hear a sweet whistling coming at you across the waters of your lake – or for that matter from the grassy banks – then you have found your wigeon; the French call them *siffleurs*, or whistlers. And its scientific name – used by both the English and the French and by anyone else in the world who has anything to do with educated conversation about ducks – is *Anas penelope*. A mallard is *Anas platyrhyncos* – not a good name for a dog.

You will find quite a lot of ducks have a first name of *Anas*: gadwall, teal, shoveler and others. But then you come on to the tufted duck, more usually called just tufty. That is *Aythya fulligula*. And there are others with the same first name, notably the pochard.

The first name is for the genus, the group of birds to which the species belongs, and it's the generic name. The second name is for the species, and is therefore the specific name. Sometimes, just to be confusing (but it isn't really, not once you've got your eye in), the two names are the same. The greylag goose is *Anser anser*, while the white-fronted goose is *Anser albifrons*.

So let's get back to the ducks: the *Anas* ducks are mostly dabblers and vegetarians, while the *Aythya* ducks dive and take mostly animals. All these ducks fit into the larger classifications of *Anatidae*, which includes geese and swans. And they're all in the still larger category, the order of *Anseriformes*. This group includes all the birds we loosely call waterfowl, with the addition of three species of screamers and the magpie goose.

And all these belong in the class of *Aves*, or birds, which is part of the phylum of *Chordates*, or backboned creatures (same as us), and they all belong to the kingdom *Animalia*, which I probably don't have to translate for you.

So there we are: from a single name to the entire structure of life on this planet in space of a single quack or whistle.

Names matter. Scientific names are essential for precision across languages and cultures and are intended to be utterly unambiguous. There is a bird we call a robin in this country, and an unrelated bird in America that's also called a robin. But their scientific names are beyond confusion: *Erithacus rubecula* and *Turdus migratorius*.

I have found scientific names useful when birding

abroad, especially in Latin America. Silvia Centrón, a great Paraguayan conservationist, spoke English better than I spoke Spanish, but only just. So we had to fall back on Latin: the bird names we were both familiar with. (I should add that Silvia's English improved radically after sundown and the first beer of the evening.)

Common names, vernacular names – like wigeon or *siffleur* – are less precise, but they are packed with meaning. They speak of the long interaction between human culture and non-human life. Some of the charming butterfly names speak of the lust of the collector – Camberwell beauty – or of the watercolourist agog to paint the lovely clouded yellow.

Plant names tend to be still richer, but I'm avoiding plants as far as possible in this book because I get them wrong alarmingly often. My old friend Ralph, mentioned before in the context of dolphins, helps me with plant names: lady's bedstraw, ragged robin, navelwort, black-eyed Susan, sneezewort, wormwood.

A vernacular name brings a creature to life. I have written about the conservation of a spider called *Attulus distinguendus*. That is less likely to touch your heart than the English-language name of 'distinguished jumping spider'. A distinguished jumping spider has a right to live: we would be more reluctant to grant that privilege to a jumble of Latin.

Every field guide to birds is full of vivid names like

skylark and raven, functional names like dunnock and wren, pedantic names like lesser spotted woodpecker, misleading names like Dartford warbler or Kentish plover, inappropriate names like garden warbler, names that celebrate humans, like Cetti's warbler and Savi's warbler, mysterious names like phalarope, names suitable for bad jokes like bustards and shags and names that have thrilled humans across countless centuries and provided a million symbols as they did so, birds like doves and eagles.

Read them, learn them, cherish them: all those wild names. Adam's first job was to name all the animals: it's a pity he wasn't told – or perhaps he just forgot – that finding the name is only the beginning of a relationship. After the name comes liking, affection, love, understanding, responsibility and a desire to cherish. We need to cherish our distinguished jumping spiders, our smews and our wigeons and our mallards.

In the great television series *The Good Life*, Tom and Barbara tried and failed to avoid the terrible trap of giving their chickens names, and always felt sad when they killed one for supper. The great outcry about the lion shot with the bow and arrow by the American dentist came about because the lion had been given a name – Cecil. How do we stop people illegally shooting hen harriers? The semi-facetious answer comes back: give 'em all names.

A name changes everything. But a name is only the beginning.

15

READING THE SECRET SIGNS

Messrs Moony, Wormtail, Padfoot and Prongs
Purveyors of Aids to Magical Mischief-Makers
are proud to present THE MARAUDER'S MAP

– *Harry Potter and the Prisoner of Azkaban*,
J. K. Rowling

I solemnly swear that I am up to no good. These are the words that activate the Marauder's Map, inexplicably given to Harry Potter by the Weasley twins. It not only shows all of Hogwarts School of Witchcraft and Wizardry and its grounds and all the secret passages, it also tells you where everyone is. The map detects the presence of Peter Pettigrew and that's crucial to the plot of *The Prisoner of Azkaban*. The map is not fooled by magical attempts to conceal identity: it sees through polyjuice potion, animagi and cloaks of invisibility.

I can offer you a spell that has a little of the same kind of magic. With a little knowledge – but more with the habit of looking – you will be able to look at the countryside and know at least something of who has been here before you. In the film, the Marauder's Map was animated with footprints that magically appeared on the parchment; in the real world you can learn to look at the countryside and read the footprints that the animals have left behind them.

The easiest are deer. Deer prints are outstandingly different from the shod prints of humans and the four-toed, four-clawed prints of dogs. Look for two side-by-side clefts in the earth that appear a little like two sugared almonds. Deer have cloven hooves, like most ruminants. They are part of the group of even-toed ungulates that includes cows, pigs, sheep, giraffes, hippos and camels. (The odd-toed ungulates include horses, rhinos and tapirs.)

So if you see the print of two toes together as you walk through a wood, you are entitled to guess that a deer has been that way. If there are a lot of prints and they're pretty big, you can guess red deer or fallow deer; small and on their own and it could be roe deer, Reeves's muntjac or Chinese water deer. If you are crossing an upland meadow, be less excited: you're more likely to have found sheep. However, if you are in the Forest of Dean or one or two other places in Britain, and the twin-toed tracks are accompanied by a lot of freshly dug earth, you've probably found the track of wild boar.

But let's say you have found the prints of deer. Probably that day, or maybe the day before, a deer walked where you are walking now – and it's a poor person who isn't the tiniest bit richer for such knowledge. It's not just the deer that imparts these riches: it's also the human pleasure of being part of a secret. You, with your sharp eyes and your still sharper mind, have penetrated the mystery of the passing deer – and no one else knows, not even the deer.

You need to find the right kind of ground: the kind that takes a footprint and holds it for a while. On the soft mud beside a pond you might find a four-toed footprint that looks big enough to belong to a velociraptor: that'll be a heron. As you start to gaze at these secret signs, you'll find that the footprint often looks too big to fit the animal you had in mind. That's because when you make a footprint you push the footprint-making material aside. Look at your own prints after you have crossed an area of soft sand: you'll swear a giant must have walked there.

I was once in a wood in Zambia that was managed – to use the term loosely – by a mining company. The executives assured us there were no mammals left in that wood: 'Ach, man, I tell you – there's nothing! Poached out, man. Gone.' I noticed a termite mound and knew that termite clay makes a superb matrix for prints. There, on a single square yard of clay, I found prints of genet, mongoose and duiker. A subsequent biological survey showed that the wood was in fact jumping with life – but it was life that has grown rather keen on keeping out of the way of humankind. Rather like most of the wild mammals on our own shores. But you can penetrate their secrets by keeping half an eye on the ground as you walk.

It's a skill, like speaking Italian – and in the same way, you don't need perfection to find the skill useful and illuminating. If you have great Italian you can read *The Divine Comedy* in the original and discuss its meaning with Italian *professori;* if you have a little basic Italian you can find your way across Florence, order *una birra* and say *ciao* to nice people as you go. With the most basic skill at reading footprints you can travel a little further into the secret life of our own country.

I have often seen master trackers at work in Africa. I remember a day I spent with Sam Manyangadze in Zimbabwe, a one-eyed veteran of the Bush War. We were to track a certain species of odd-toed ungulate. 'What are our chances, Sam?'

'If you will walk all day, 100 per cent.'

'Then tomorrow we will see rhinos.'

We walked for six and a half hours. Not a gentle stroll, stopping to look at nice birds and to have a cup of tea beneath a shade tree: this was a fierce yomp, with Sam's one good eye picking out the signs – of which the one that mattered most was the three-toed tracks of the black rhino's four great paws. Eventually we came upon him . . . and got to within twenty paces of the wondrous beast.

In the Luangwa Valley in Zambia during the dry season, the sandy bottoms of the dried riverbeds are as full of news as a local newspaper. With the help of experts, I have tried to work out which direction the elephants were travelling, and whether that was a hyena – look for the claw-marks – or a leopard, and exactly how many lions passed that way in the night.

When I get back to Britain, I have never felt disappointed at the shortage of elephants and of odd-toed ungulates with horns on their noses, or because the tracks of non-human, non-domestic animals can be hard to find. Rather, the savannah has tuned me in. As a result, I am better at noticing the tracks you can find in this country.

You can tune in without going to Africa. The trick is to cultivate a certain curiosity about marks on the ground. You notice, you look, you examine, you make your guess. You can take a picture of the print, establish its size with a coin or some other object, and check it out when you get home: confirmation is a few clicks away. As a result, you are that little bit closer to the wild creatures you share the place with.

Then you get the treat of a fresh fall of snow. Sometimes, if you get there before other humans and their dogs have disturbed it, you can find yourself gazing at a complexity of tracks that's almost as mad as those in the Luangwa Valley.

I remember just such a day in Suffolk. There was a footpath: it ran along a hedge past a playing field, then alongside two or three arable fields. Nothing special. The path was much used up to the first stile, but just about everybody turned back at that point. By the time I had crossed the second stile, there were no prints of humans at all . . . and yet the field before me was a dazzling mass of prints. All that life, going on outside human sight, beyond human knowledge – it would have been thrilling even if I'd have looked no closer. But, of course, I did.

There were plenty of deer prints, all muntjac, I guessed; I had never seen roe deer here and it was the wrong habitat – not enough water – for Chinese water deer. Also there were a lot of rabbits about the place: that long, broad hind-foot is a giveaway. And that made it all the easier to work when I had found a hare: the same basic idea but a very great deal bigger. It was then that I noticed a slim set of doggy prints, but even as I dropped to my knees – waterproof trousers! – to have a closer look I realised that no dog had been here since the snow fell, for there were no human footprints. Besides, these were slim paws, the marks more or less in a straight line, rather than straddling from side to side. A fox, then. And, as I followed the prints a little way, I found that they intersected with the hare's prints, and that there were suddenly huge gaps between one set of hare prints and the next. The fox had startled the hare, hunkered down out of the wind, and the hare had exploded into action with the suddenness that hares can manage at times of stress. I could see that the fox had followed for a few yards and then, being a smart animal, realised the futility of further chase. Can't compete with that.

Here, then, was a small drama, written out for me on the great white page of the snow, in careful script that was there to be read by any who could, in the manner of the Rosetta Stone. Here was the story of the hungry fox and the hare that lived.

As you begin to get a little more aware of these signs on the ground, you realise that there are many creatures living among us who are seldom, if ever, seen. You can now activate the spell that sees through their cloaks of invisibility. You may find some scratchy-looking prints on a flowerbed in your own garden, almost like a tiny hand: that's likely to be a hedgehog. Or, on a riverside walk, you might take a closer look at the doggy print – quite a small dog – and see that it's got five toes, and they're joined together. They're webbed. And there's an otter. Nod wisely to the flowing water alongside: nice one, Otter, you can say. You didn't fool me, but don't you worry about that. I won't tell anyone. You're secret is safe with me.

And then, like Harry, you can deactivate the Marauder's Map: 'Mischief managed.'

16

A Vision Seldom Seen

They could see more light than they had ever seen before. And the deck and the sail and their own faces and bodies became brighter and brighter and every rope shone. And the next morning, when the sun rose, now five or six times its old size, they stared hard into it and could see the very feathers of the birds that came flying from it.

– *The Voyage of the Dawn Treader*, C. S. Lewis

I have been suggesting that you improve your walks (and your sits) by looking for footprints, for turds, for secret paths, for scrapes and scrabblings in the earth, for disturbed vegetation. Am I crazy? If you follow all this advice, aren't you going to fall down, bump into things, wander off the cliff path and go tumbling into the waves beneath?

But no. You can look for all the things I suggest while maintaining perfect balance and direction and without putting your life at risk even a little bit. You can also see birds, butterflies, dragonflies, all kinds of living things *without actually looking at them.*

It can look like magic and it can feel like magic – and it's a piece of magic we're losing as we become less wild in our lives. It's something we can appreciate most fully when walking in the bush in Africa. Out there you are very much more aware of everything all around you than you are normally. And so when you catch something from the tail of your eye you turn your head and look – not just for interest, but also because the act of walking in the bush has reactivated a survival mechanism that lies deep within you.

This same kind of hyper-awareness can be cultivated

back home. All you need to do is to be aware of what's happening on the fringes of your vision – and to respond.

Broadly speaking, we have not one but two forms of vision: two ways of looking at the world. You can call this your wild vision and your tame vision. Or perhaps your unconscious vision and your conscious vision. Taken separately: you have your front-on, straight-ahead staring vision – the sort you're using to read this page – and you have your what-was-that-did-I-really-see-that-or-did-I-imagine-it peripheral vision. And the fact is that we are losing this second kind of seeing.

But we can get it back.

Sir Clive Woodward was an exceptional rugby coach, and he led England to victory in the World Cup of 2003. His attention to minor details won him both admiration and mockery. Among dietary experts and fitness coaches and separate experts on defence and attack, he also took on Sherylle Calder as visual awareness coach.

She has since worked with other rugby teams, with those in golf, motor-racing and American football, and is now working again with the England rugby team. The main thing she deals with is peripheral vision. You can find her work on her EyeGym website. 'In the modern world the ability of players to have good awareness is deteriorating by the nature of mobile phones,' she said.

That and the other screens we look at in the course of our day: hours and hours in which the only thing that matters is the thing directly in front of us. As a result, our brains are now less likely to respond to stuff at the far edges of our vision. And if you want to be aware of wildlife – either to keep yourself safe in a hostile world, or to enrich yourself with knowledge of the creatures we share our planet with – you need your peripheral vision.

We are primates. We have two forward-facing eyes that give us, in the centre of our vision, excellent ability to see depth and colour: very useful abilities when you are leaping from tree to tree, as our ancestors did. But while the centre of our vision – the macula – is essential to us, we need the rest of our vision as well.

Here's an experiment to prove it. Get up, close your eyes and stand on one leg. Whoops! It turns out that you need to be able to see in order to stand still and balanced. The information that comes from the edge of your vision allows you to be aware of your surroundings, so you can make minute, almost imperceptible adjustments of your position and pre-empt the major wobble you experienced with your eyes closed.

Your peripheral vision comes in handy in many urban circumstances, like walking across Liverpool Street Station in the rush hour, or crossing the road and keeping safe from rogue cyclists. Loss of peripheral vision is a serious and damaging condition: tunnel vision – a term now used figuratively in many different circumstances – is a dangerous and damaging disability.

But the way we live in the twenty-first century encourages us to concentrate on this straight-ahead staring field of vision and to value its input more than ever before – and to do so at the expense of the other kind of seeing. Our brains are getting a little bit lazy when it comes to the messages that come from the edges.

You learn about peripheral vision if you spend time with prey animals. Horses' eyes are set to the side of their heads. They can see dead ahead with two eyes, though only in a very narrow field. But they can see all around them: more or less 360 degrees, on the vertical as well as the horizontal plane. My horse can see me when I'm standing a few yards behind her. She can also see me when I'm sitting on her back, and at the same time she is aware of the changing patterns of the ground beneath her feet. A horse can check its stride over difficult ground, seeing clearly the way ahead while also being aware of the cars, the tractor, the two walkers and the loose dog, and of you on top doing the riding.

Our own peripheral vision is not in that class – it's not supposed to be – but it still has more scope than we'd easily guess. Here's another experiment: sit still and look ahead in a steady gaze, yes, as if you were watching telly or gazing into the eyes of your beloved. Now hold your hands up, like they do in a Western, at about eye level. Move them back towards your shoulders, while still staring forward. You will find that you're easily picking up the movements of your hands at 90 degrees, and if you try a little harder, you can still glimpse them even at about 110 degrees. Now hold your hands above your head and below your chin: and without moving your head, you will be aware of them, especially when you waggle your fingers.

So yes: you have more visual scope than you bargained for, and the fringes of your vision are particularly good at picking up movement. In fact, such information gets to your brain about 25 per cent faster than messages from the centre of your vision – and does so because it's essential for survival. Peripheral vision gives you balance and helps with your movement. It also gives you warning of moving things.

Peripheral vision has its own circuit in the brain, and it's one that deteriorates with dismaying briskness in people with Alzheimer's. Such people lose physical confidence and fall over more readily and more often. It follows that maintaining your peripheral vision has plenty of benefits. Even those who never spend a second thinking about wildlife need to cultivate peripheral vision.

You can work at those hands-up exercises, and no doubt that will help. Athletes – young people in their prime – are frequently set routines to improve their peripheral vision: awareness of space is crucial in all sports. Try this: take an object – a tennis ball, say, or a sealed bag of sweets – and toss it above your head while maintaining that same implacable still-headed stare straight ahead. The idea is to catch the falling object at about waist-level without moving your head. It seems amazing that you can even get close, but I catch the object more or less every time. It feels refreshingly like magic: try it with an egg if you're feeling brave. It's the way jugglers work: that fixed gaze is all about the core of their art, which is throwing. The catching takes place on the fringes of their vision, and with as little hand movement as possible.

Your task, then, is to take this form of enhanced visual awareness out into the wild world. The key is simplicity itself: indulge your peripheral vision. When you see a fleeting shadow at the corner of your eye, give into it and turn your head. When you think you might have glimpsed something in the sky above, stop and look up, because you probably have. When you are half-aware of something at your feet or on the side of the path, turn your head and give it your full mind: it might be a badger's snuffle-pit. Make a point of doing that every time and you start to find wonders. I have been the last of a group walking in single file along a narrow path, but it was me that picked out the slow worm and brought it to everyone's attention.

Many people are better than me at such miracles – and once again, many of the people I have walked with in Africa make me feel like a blind man. But I've got better the more I've looked at – and for – wildlife. I know that a flicker in my left eye might just be the best moment of the day: so my head has turned long before the idea of doing so has reached the level of conscious thought.

And that's how I tend to think of peripheral vision: the vision of the unconscious mind. It's about stuff you're aware of without knowing of your own awareness. Thus we walk across Liverpool Street Station without getting shoulder-charged to the ground, or we hang back on the level crossing just at the moment when the cyclist decided that pedestrians are beneath his notice – just as our ancestors were aware that the lone buffalo had got to his feet and was considering aggressive action.

The more we filter out these messages, the more we lose the wild world. The more we respond to them, the wilder our world becomes. Where once I saw a shadow I now see a stoat. That shift of light above your head in a wood: always stop to look, because it might be a treecreeper. It's almost always pure movement you get from the corner of your eye, but as you turn your head to meet it you bring a different part of your eyes (and your brain) into the action and the movement becomes filled with colour and is glowing with a glorious yellow: yes, that's a brimstone, a favourite butterfly, perhaps the first one of the year, and you have a spring in your step you would never have had if you'd kept your head still.

Out of sight, out of mind. But your sight is bigger than you suspected. And so is your mind.

17

REGAINING YOUR LOST SENSE

In the darkness something was happening at last. A voice had begun to sing. It was very far away and Digory found it hard to decide from what direction it was coming. Sometimes it seemed to come from all directions at once. Sometimes he almost thought it was coming out of the earth beneath them. Its lower notes were deep enough to be the voice of the earth herself. There were no words. There was hardly even a tune. But it was, beyond comparison, the most beautiful noise he had ever heard. It was so beautiful he could hardly bear it.

– *The Magician's Nephew,* C. S. Lewis

The last chapter was about the loss of half a sense. This chapter is about the loss of an entire sense – and how to get it back again. We talk about people who lose their hearing and feel sorry for them. Alas, the fact is that we are all losing our listening. We have all mastered the art of tuning out. We have done so not wisely but too well.

Life in cities would be impossible if we didn't. If we didn't edit out the roar of traffic we would be unable to cope. When I lived in the middle of Hong Kong, I remember travelling to an outlying island and feeling the silence like a physical blow, one so confusing to my ears that it seemed to affect my sense of balance.

We have developed a horror of silence. There is hardly a pub or a restaurant that doesn't play music all the time: music we are not supposed to notice; it's just wallpaper, designed to cover up the awfulness of silence. It's there when I go to the bloody Co-op to buy eggs. We'd go mad if we let all these noises impinge on us, so we have pulled off a clever mental trick. The sound is still there, but it no longer touches us, or even reaches us.

We have become so efficient at this that the sounds of the wild world hardly get through at all. Well-researched science proves beyond all doubt that the sounds of nature are good for us in all kinds of calming, enriching and fulfilling ways. But even when there is birdsong all around us, we can't hear it. We no longer hear the blackbird's song coming down from the chimneypots, just as we no longer hear 'Sex & Drugs & Rock & Roll' when we're standing in the queue at the supermarket till.

We have always been a species convinced that sight is the supreme sense, the sense above all others. That truth is expressed in the way we speak: if you see what mean; well, look here, if you want my view, it's something we should all keep our eyes on . . .

And this habit of thinking is part of the way we understand nature. We go bird*watching*, as if the experience of a single sense was all that counted. If I've been out for a good walk, I might be asked: 'Did you see any good birds?' In spring especially, a better question is: 'Did you hear any good birds?'

Plenty of birders won't count a bird – won't put it on their list, won't acknowledge the experience of identifying the bird – unless they have set eyes on it. They feel that counting a bird you've 'only' heard is cheating: a sin against the strict moralities of bird*watching*. And it's a great nonsense. I would have the greatest difficulty in distinguishing a willow warbler from a chiffchaff if I saw one, and even with an exceptionally good view I wouldn't be 100 per cent confident. There are plenty of stories about identification experts with the birds in their hands – caught for ringing – who have confused the two species. But I can tell one from another with complete confidence when they sing, and it's dead easy. It's presumably how the birds tell each other apart, so it's foolish not to allow them to help us.

Birdsong. Everybody claims to love the stuff. A city park, where you can hear the birds above the sound of the traffic, is a treasured place. All the more reason, then, to tune into it.

So I invite you to read the next couple of pages with special diligence – because it might just be one of the most significant things you ever do. I'm not going to give you a sonic identification guide to every British bird, or even a beginner's guide to the most common. (I've already done that in a previous book called *Birdwatching with your Eyes Closed*.) What I'm offering here is a starter-pack. Once you've got that, you can find other ways of learning more. You can get very good (and very cheap) apps on your smartphone that will give you most of the common bird sounds at the touch of a button.

My starter-pack works on the same principle as the buddleia bush, and it comes down to names. And in the usual way, this is not about learning names to show off or to advance the cause of science, but to bring you into a state of greater intimacy with the creatures concerned – and with the rest of the wild world.

So here are five birdsongs. Song is the way some birds claim and hold a territory; a territory is an area in which the birds concerned can claim a mate, build a nest, hatch eggs and raise their young to maturity. Song is mostly – in most species exclusively – a male thing, but it all comes down to female choice. The males sing, the females choose the singer they like best. A song is not a small thing. It's a huge physical effort and it expresses – and it's supposed to express – important truths about the bird's fitness, strength, maturity and experience.

Song is the first line of defence when it comes to holding a territory, and it's usually enough. But early in the season, a song may need to be backed up with display: posturing, showing how great a bird you happen to be. This can escalate to physical contact, and – though rarely – to full-on violence.

Birds make noises for other reasons too: to communicate and express danger, or to keep in touch with each other. And while that's all very intriguing and I hope you'll come to such things in the fullness of time, I'm going to ignore them. Instead, here are five songs sung by five common birds: birds you'll find in most gardens and parks. Once you've learned to pick out their voices from the rest, you can grow more ambitious. So let's crack on.

1. Robin

This is a good starting point because, more or less uniquely in this country, the robin sings for most of the year. They hold territory in the winter and defend the food resources within it by means of song. So if you hear a song of any real complexity on a winter's day, it's almost certainly a robin.

It's a pretty song: thin and musical. It's been called wistful, even sad, though I don't see that myself. It's especially moving in these cold months: a lone voice rising above the forbidding landscape, determined that life should continue. If you make a New Year's resolution to learn birdsong, by mid-January you should have robin sorted.

2. Great tit

Great tits make a lot of noises. Never mind. Let them. You are to wait until they make one special noise. They do it to announce the start of spring, and I've heard it as early as December. Great tits are birds of tremendous enthusiasm and they can't wait to get on with the job of making more great tits. The most frequent refrain when a male great tit gets down to business is as simple a song as you could get: two syllables, endlessly repeated.

Teacher! Teacher! Teacher! That's just one of the ways the call has been transcribed: two syllables with an immense stress on the first. It's also been described as a leaky pump. It's loud, simple and very strong. Get it into your head and you'll hear it throughout the spring (and perhaps you'll thank your teacher-teacher).

3. Wren

Wrens are one of the smallest birds in Britain and one of the loudest. They are birds of low bushes and tangles of brambles, so if you hear a very loud song from about knee-high, chances are you've found a wren. But there's a way of making absolutely certain. Wrens chuck out a number of hurried notes at you, gabbling in their haste to get the song out. And then, at the end, they give a prolonged and violent trill.

So listen out for that moment of confirmation: when the wren gives a trill. They almost always do this from cover, but when you see a wren giving out a trill, you might think the poor bird would sing itself to bits in the violence of its song. Hear your trill and you've found your wren.

4. Song thrush

A song thrush loves repetition above all else. It takes a short phrase, sings it two, three or four times, and then has a short pause before taking up another phrase, repeating it in its turn. The song is loud and likely to come from the top of a tree. A song thrush loves a good song-post and will move from one to the next around its territory, singing and repeating.

It's a good deal more musical than it sounds. And it's tremendously variable: a song thrush will take sounds it likes from all around – a nuthatch, a tractor reversing, a lawn-mower – and turn them into music. Like most musicians, song thrushes like to take a theme, have fun with it and then repeat it, and then move onto something new. If you're hearing musical repetition, you're listening to a song thrush.

5. Blackbird

I have put these five birds in the rough order in which they start to sing in the advancing year. You'll get plenty of variations in that, but you can be more or less certain that the blackbird will be last of these five. And the blackbird is the most melodious of these five singers – or perhaps of any singer in this country.

The song is a simple, gorgeous laid-back fluting, as if the blackbird was leaning against one of those chimney pots with his hands in his pockets, whistling. The song is urgent for the blackbird – life and death, nothing less – but to us listening humans it's the most relaxing song of them all. It tends to be heard on warm evenings in spring, the first evening we dare to take a drink outside.

If it's the sweetest song you ever heard, it's probably a blackbird.

* * *

Now please note that all these are resident British birds. They spend the winter with us and so they are here and all ready to sing when the time comes. They have the jump on the migrants: by the time these migrants are in song the entire country is singing its head off. May is the best time to listen to birdsong, but the worst for trying to learn it, as if you were trying to learn the instruments of the orchestra by listening to the *Ring Cycle*.

But believe me, when you have advanced a little in your study of bird music, the month of May will fill your heart and soul. You will be able to walk through a wood seeing not a single bird, but knowing every bird that is singing up in the hidden choir-loft of the canopy. You will be calling them down as if you were Radagast, the magician of the birds in *Lord of the Rings*.

Knowledge of song will improve your performance of the bottomless sit. There is all the difference in the world between listening to music you love and noise that's imposed on you in a supermarket. In the same way, you will no longer be listening to a vague thing called 'birdsong': you will be listening, well, to music you know and love. You listen, you become part of it.

As you acquire the habit of listening you begin to hear other sounds too: the barking of muntjac, a splashing in the water that might be a duck or might be an otter, a plop that might be a water vole or a kingfisher. When you try that travel-through-time spell and get up at dawn, you will be all the richer for the singers you manage to identify.

And perhaps the richest thing of all is that once you have tuned in to birdsong, you will never tune out. You will never be off duty. Nature will be with you always. I remember being baffled by a survey that asked how often I went birdwatching. I don't go birdwatching. I *am* birdwatching. Or bird-listening. Or just birding. Or just wildlifing. Either way, I am infinitely more in tune with the wild world than I was before I began to learn the songs and the calls, and learning birdsong was the biggest single step I made on that journey.

So you can be walking through London and still be aware that a blackbird is singing in Soho Square, or that there's a wren in the garden on the way to the station. You don't need your bins, you don't even need to turn your head, and yet your day is enriched. Tuned in. Revelling in a lost sense.

18

THE MAGICIAN'S LIBRARY

Why, if we can get back to our own world by jumping into *this* pool, mightn't we get somewhere else by jumping into one of the others? Supposing there was a world at the bottom of every pool!

– *The Magician's Nephew,* C. S. Lewis

J. K. Rowling has always loved this chapter in *The Magician's Nephew,* entitled 'The Wood Between the Worlds'. 'It's a library!' she said. Which is a beautiful idea: every volume you open will take you to a new world, if you only let it. There is magic of a kind in every book that was ever opened.

So let me tell you about my old friend Manny Mvula. Manny was the first black Zambian to get his full guiding licence in the Luangwa Valley. The process involved a rather frightening interview, in which Manny had to face a panel of rather difficult old-fangled white men who had lived in the Valley all their lives and loved it with a wild passion.

The more questions they asked, the more Manny dazzled: scientific names of trees, seed propagation, the social system of lions, the evolution of hippos, the brood-parasitism nature of indigo birds. Eventually one of them cracked. 'Mr Mvula, how come you know so much?'

Manny smiled kindly. 'Well, I have read a lot of books.'

As you move deeper into the wild world, you will want to understand more, and books are an essential part of that process. Not just field guides and other books that help you with identification – yes, by all means get hold of these, as many as possible, but improving your ID skills is only a small part of rewilding yourself.

You don't just want wilder eyes and wilder ears: you also want a wilder heart and a wilder mind. As always, the right books are there to enrich you. So here are ten favourites of mine, in no particular order.

1. *The Jungle Book* by Rudyard Kipling

Please note that the book has nothing whatsoever to do with the film. Kipling's animal stories are variously interpreted as allegories about colonialism, but the thing about Kipling is that he was never able to stick to a pre-planned moral purpose. He was too good an artist: the story and the characters take over. ('Never trust the artist; trust the tale.' – D. H. Lawrence) So we have Mowgli in the jungle on terms of intimacy and equality with the creatures he shares it with, and that is where the magic of these tales comes from. The jungle itself is the main character. There's more than the Mowgli stories: 'The White Seal' is a story about a seal's search for a safe beach far from humankind: as vivid – and as lovely – a tale of the ecological holocaust as has ever been written.

2. Haiku by Basho

Haiku is a traditional form of Japanese verse: ultra-short – seventeen syllables in the original – and often revealing a single instant of nature.

Sparrows in eaves
Mice in ceiling –
Celestial music

And you either get it or you don't. As said before, wildlife is full of instant dramas, played out to an audience of one, but just occasionally they can be caught in seventeen syllables and turned into a half-lost memory of your own. There are many collections of haiku, and there are other masters who can bring moments as vivid even as Basho's – though Basho always has more. A wild moment, three hundred years gone, and yet here it is living again in a flashbulb moment of understanding: a personal telegram from the wilds of the seventeenth century.

3. *Birds Britannica* by Mark Cocker and Richard Mabey

Not a field guide but a collection of facts and anecdotes about each species of bird on the British list and how it has affected human lives. You can find a picture of a stone curlew nesting in a herbaceous border, a history of bird of prey persecution, the superstitions that surround robins, blue tits and their period of fascination with milk bottles, and a label for Woodpecker cider. It's a treasury of information: telling us how much birds mean to us and, often enough, how poorly we humans have kept our side of the deal.

4. *Ever Since Darwin* by Stephen Jay Gould

As you get wilder, you need to understand how it all works. How it's all put together. That means understanding evolution – and it's deeply shocking. We cherish the myth that the whole point and purpose of evolution was to produce wonderful loveable us. Then we learn the truth: that this is not the case at all. Evolution is not about goals or the quest for perfection: it's a chance-driven process that's all about muddling through life and eventually becoming an ancestor. Or not. Our place in the world is not the result of our perfection but the chance impact of a meteor 65 million years ago – the one that did it for the dinosaurs. (Dinosaurs were around for 100 million years: we have a way to go before we can match them.) These truths demand to be understood. Gould's early essays are witty, elegant, concise – and occasionally jaw-dropping.

5. *The Butterfly Isles* by Patrick Barkham

This is a quest book, and a good one. Barkham sets off to see all fifty-nine species of butterflies that breed in Britain, and he rewilds himself as he goes. It's worth reading no matter what your level of expertise, but for butterfly-beginners his ability to capture the nature – the vibes, the character – of each species is very pleasing. The process of personal rewilding is above all about love, and this book, in its understated way, is full of the stuff.

6. *Beyond Words* by Carl Safina

As you get wilder, you find some of the old certainties slipping. Is life really about us humans on the one side, and every single other living thing somewhere else? Doubt that, even for a second, and you find yourself sliding into the bandit country that lies between humanity and everything else. Safina spends time with some of our fellow mammals, choosing the most famously social species: African elephants, wolves and orcas (or killer whales). As he writes you find the adamantine barriers we have built up between ourselves and everything else that lives are not as strong as we thought.

7. *My Family and Other Animals* by Gerald Durrell

I've never been mad about high and solemn nature-writing – the sort of thing that's been described to me as 'Writing with a capital R'. In this glorious book, Durrell writes of an Arcadian childhood on Corfu and his boyhood fascination with the wildlife of the island, all mixed up with the doings of his own eccentric family. And no, it's not an awful lot like the television series *The Durrells.* The author can switch from humour to gorgeous description in an instant: his birthday expedition in his new boat, *The Bootle-Bumtrinket,* is perfection. This is a true classic of real – and natural – nature-writing, not least because it's about joy.

8. Poems by Gerard Manley Hopkins

I'd put Hopkins at the top when it comes to nature poets writing in English. There are many good things to discover, with kingfishers catching fire and a great anthem for wetland conservation. His best-known work is 'The Windhover', a poem about a kestrel that captures perfectly that sense of instantaneous glory and perfection that can strike without warning at any time to anyone who is sufficiently wild.

> *My heart in hiding*
> *Stirred for a bird, – the achieve of, the mastery of*
> *the thing!*

9. *The Diversity of Life* by Edward O. Wilson

Life works by making lots and lots of different species. It's not so that we can collect them or make lists of them – even though that's a gratifying process to many people – rather, it's the basic mechanism of life. Wilson, a brilliant scientist and a great conveyor of ideas, gives you a guide to life and the way it operates. He suggests that the more we lose diversity, the closer we get to destroying life – all life – as it is lived on earth today. Some say the loss of biodiversity is a greater threat to the future of the planet than climate change. Perhaps, he says, that's a wrong view. He then adds the most chilling line in scientific literature: 'One planet. One experiment.'

10. *The Cultural Lives of Whales and Dolphins* by Hal Whitehead and Luke Rendell

A few years ago, an academic who made even the suggestion that non-human species could pass on notions by means of cultural transmission – just like us humans – would have been kicked out of the profession. It was heresy. But it's also the plain truth. This fascinating book tells us truths about the variations in the songs of humpback whales – cultural shifts, changes in taste, nothing less – and how a certain population of dolphins off the coast of Australia – but no other – have developed the practice of using sponges to project their noses when foraging at the bottom of the sea. Life has greater possibilities than you ever dreamed.

* * *

In addition to these ten books, since *The Chronicles of Narnia* gave me the notion for this book and its stories haunt these pages, you could always give those seven beautifully turned fairy tales a read, or a reread. It's become fashionable to knock the *Chronicles* these days because the books – though they are many other things as well – are Christian allegories. But they're not Salvation Army pamphlets: the books are a series of wonderful – and magical – stories. As with Kipling, the tales outrun the writer's conscious intention.

Throughout the pages of the *Chronicles* non-human animals walk and talk and perform mighty deeds: like Reepicheep, the heroic mouse, or Fledge, the first of all winged horses, or Jewel the unicorn in the last days of Narnia. The human characters treat the non-human characters with respect and understanding: on terms of equality. The spell of that bit of magic has never truly left me.

I've put together a funky old collection of books for you. They will bring you the joy of a child, the thrill of adult understanding, a little real science, the piercing insights of poetry: and together, all of them – and many others too – will convince you that there aren't really two worlds out there, one for humans and one for everything else. We're all in it together.

19

How to Turn into a Swan

He hardly heard what Professor McGonagall was telling him about animagi (wizards who could transform at will into animals), and wasn't even watching when she transformed herself in front of their eyes into a tabby cat with spectacle markings around her eyes.

– *Harry Potter and the Prisoner of Azkaban*, J. K. Rowling

I've tried being an eagle, and I have to say it's pretty good. But the spell is a mite complicated and beyond the reach of many of us: it involves riding pillion on a microlight, which is a hang-glider powered by a sort of lawn-mower engine. It's best done over some wild place, and the thrill of being a gigantic bird is not something you ever forget.

But it's easy as anything to turn into a swan, and that's the spell I'm offering you in this chapter. You too can travel the river with calm unhurried majesty: you too can become part of the river itself. What you need is a canoe.

There are many different sorts of canoe and they all have their points. What I'm talking about here is the Canadian canoe. In its non-sporting form, it's a rather portly craft and you propel it by means of a single-bladed paddle.

We get the idea of the canoe mixed up with white water and the Eskimo roll: enthusiasts for such stuff are never happier than when inverted. We think that canoeing has to be thrilling and spilling and intrepid, that you're mostly soaked through and agog for the next capsize. And while such things are no doubt lovely, it's not my idea of canoeing.

A Canadian is not very dashing. What it's good for is

pootling. You're not rushing about or dicing with death: you're sitting on the water, moving along at your own pace with a series of firm but untaxing strokes of your paddle: just like a swan, in fact.

We should at this stage reject the popular notion of the swan's movement: serene on top while paddling furiously out of sight. Swans are perfectly capable of doing things furiously, as many an angler will tell you, but not paddling. Frenzy is far from their minds: it's a firm, controlled one-two, with alternating strokes of those enormous black paws. The one exception is when a male swan wants to impress you – more often another swan or even a goose that has strayed into his orbit – and with wings raised and neck back, he will scoot towards his target with ferocious double-strikes of the feet. The display is called busking; it's highly impressive and you get a good chance to witness it when you're out on the river and paddling away yourself.

But, you will point out, I don't have a canoe, and I'm not about to buy one no matter how much you go on about swans. That is a mere detail. Many places along rivers and lakes hire canoes by the hour for a very reasonable fee. Nor does it take any great mastery to make it go. Even the mysteries of steering it are straightforward enough. All that matters is getting out on the water.

The opening few minutes of a canoeing session for an inexperienced paddler will be dominated by the task of driving the boat, but this is a phase you go through soon enough. Once you've got your rhythm, you forget about the mechanics and you begin to pick up your instructions from the river itself.

It's usual to have two people in the boat, but you can make a go of it on your own, as I have done many times. If the boat is long enough, you can even have three. You can also find room for luggage, children and dogs. The paddler at the back does most of the steering. Here is the one bit of technical advice you need: the rear paddler mostly controls the direction of the boat by means of a J-stroke. This is just what it sounds like – a stroke with a hook at the end: pull the paddle towards you through the water and then finish with a sideways flick away from you. Even when you're on your own you can keep the canoe in the right direction without swapping your paddle from one side to the other: a good thing to avoid as this generally involves decanting half a pint of river water into your crotch as you make the switch.

Once you've got your J-stroke going, you can stop worrying about hitting things and stop laughing about your ineptitude and start being a swan. The magic will start working. Your mind adjusts to the speed of your progress: which is mostly slow. Your understanding of landscape and waterscape is very closely related to the speed you travel at. The faster you go the more big things you see. The slower you go, the more detail. At the speed of a

Canadian canoe at full pootle you are experiencing the water with an intimacy and an intricacy denied to people rushing past on a stink-boat.

I remember my first moment of realisation: that swan moment when I stopped thinking *Ooh, this is a bit of a lark,* and started paddling properly, started seeing the details, started to become part of the river. I was at the back of the boat, my wife in the front, younger son and dog in the middle. From nowhere I remembered Modesty Blaise making an escape from a tropical island in a canoe and adopting a J-stroke. So I experimented: a stroke on the right-hand side pushed the front end left: put a hook into the bottom of the stroke and the front end corrects itself and comes back straight. At a stroke – the mot juste – the canoe was taking care of itself. Steering was no longer a matter of constant messing about. I could delegate the task to my unconscious mind and let the rest of me relax and enjoy the river.

And as I did so I started to become aware that I was paddling through a ballroom. The river was a great wet winding dance floor: the banks lined with dancers. They danced together in crowds and clouds: shining out in a deep blue that was almost black. Each individual dancer was lovely, but it was the unending crowds of them that tipped things over: all around us a glorious bioabundance of beauty.

These were banded demoiselles: like dragonflies but

slightly different. They love the banks of slow-flowing rivers and in the late spring and early summer they step out in numbers and dance: males seeking to attract a female. What a choice a female is faced with: an entire river full of gorgeousness. I stepped into that canoe for a bit of fun: I stepped out a person wholly committed to the Swan's Way.

You can buy a canoe for less than £300, or you can get an inflatable version for not much more than £100. I now have a canoe of my own and it's given me a dizzying amount of pleasure, quite disproportionate to the money it cost. It sits on the water, solid and sensible and almost as wide as Gerald Durrell's beloved *Bootle-Bumtrinket*. It's made of lovely plastic and I have yet to spend a single second on maintenance. It's reassuringly stable in most circumstances. It gets upset in high winds, because it sits on the top of the water as a flat-bottomed craft should, so a good gust will send you off course and a headwind will require a little effort to get through. The answer to that is to choose a calm day.

I have seen plenty of lovely things from the canoe: kingfishers, marsh harriers, and once a grass snake swimming across the river. Yet it's not the seeing but the being that matters. I don't even take binoculars on the canoe: my hands are full of paddle and to stop and peer would break the rhythm.

It was late May. Ralph was up for a visit, so we took the

canoe to Rockland Broad. We're old hands at the canoe these days: an easy transition from land to water and we were away, with the reeds reaching high above. We've known each other for half a century: we met when we were at school and we are at ease in conversation, equally at ease in silence, and that's a great asset on such an expedition. We crossed the open expanse of the broad and then came to a set of notices: no navigation beyond this point. But we were no stink-boat. We had no propeller to snag in the weeds, and no deep keel to catch in the mud. Where the swan goes, we go.

So we slipped between the piles that marked the extent of the stink-boat's domain and entered the soft marshy backwaters of the broad. We made no sound but the cloop of the paddles. In the stillness rose the voices of reed warbler, sedge warbler, Cetti's warbler, reed bunting, willow warbler: all for us.

At one point a notice barred our way: no admittance to the Ted Ellis nature reserve for rogue canoes. Fair enough. We wandered on in another direction, cap knocked askew by occasional willows dropping over the water, going across patches of duckweed that looked like meadows. We left a long path through the green that slowly closed up behind us. A swan glared suspiciously as we passed: it seems the fellow feeling is a one-way street. We gave him nearly as much respect and room as he wanted, the best we could do, anyway.

We found another channel. No noticeboard this time. Not much room, either. But there was surely clear water beyond the clumps of reeds that blocked our way. Could we get through? Was it a go? Well, give it a lash: we slid through, bottom softly mudding out and then releasing, paddles held inboard.

Now we were on another waterway, still lovelier, and the banded demoiselles danced their courtship dances, and I noticed that their wings were coloured with the same Quink blue-black ink that Ralph and I were supposed to use at school. We were in one of those secret gardens, one full of water, with the quiet green of the reeds set off by the loud yellows of lilies, marsh marigolds and flags.

Then, impossibly, a landing stage. One that didn't bear the ubiquitous Broadland noticeboards reading 'No Landing'. So we landed and stretched legs and let this marvellous place soak into us until we felt as if we had been there for ever.

The idyll was broken by a young fellow in a green polo-shirt. Not a good sign, that shirt. It looked official. He asked us politely what we were doing. There wasn't really a good answer to that. That good old schoolday fallback, 'Nothing, sir', was accurate but unsuitable for the occasion. 'Saving this man from drowning' was better, but unconvincing. We explained that we had just stopped. Polo-shirt politely suggested that it would be a good idea if we just moved.

It seemed we were on the Ted Ellis Reserve after all. We tried to explain that we hadn't defied the noticeboards: we had taken an unmarked channel – honestly, sir, don't give us detention. We were politely disbelieved, and we took to our canoe, apologising as we did so. They are right not to want these waterways disturbed: they belong to dabchicks and demoiselles, not to us.

We paddled away in mild embarrassment, wondering if it was time for a pint at the New Inn. Yes it was – or would be after one more lap of Rockland Broad. We looked back over the sweet secret acres that are for ever Ted: a hobby, the knife-winged falcon of the summer, fizzed overhead. Some bird. Some place.

It was on just such a day that Charles Dodgson – Lewis Carroll – first improvised a story for Alice Liddell, the story of Alice's Adventures Underground. He later wrote it down and published it as *Alice's Adventures in Wonderland*. The book begins in a boat:

All in the golden afternoon
Full leisurely we glide . . .

Wonderland is perhaps the only possible destination on such day, in such a craft.

20

How to Walk

Come in by the gold gates or not at all,
Take of my fruit for others or forbear.
For those who steal or those who climb my wall
Shall find their heart's desire and find despair.

– *The Magician's Nephew,* C. S. Lewis

It's generally accepted that nature is good for you. There are millions of stats and experiments showing us that being out in nature makes us happier, healthier, better tempered, more fulfilled, better able to work well, better able to have fun, able to live longer and live better, to recover better from misfortune, better able to enjoy life and better able to endure it.

But it's no good going out into a wild place and saying: 'Well, come on then, nature, start doing me good. I'm ready for you: make my life better.' In a way it's like marriage, or like a good marriage. There's no doubt that a good marriage is good for you, but you don't get married because you've read that marriage will do you the world of good. No: most of us get married for love. For love not of marriage but of another person. So if you want to improve your life by means of nature – a very reasonable and easily attainable ambition – the person who goes out asking 'what's in it for me?' will find the least good and the smallest benefit.

Don't seek health; seek wrens. Don't go looking for eternal life; go looking for birds. Don't set out on the track of wellness and mindfulness; set out on the track of deer.

Don't dedicate yourself to a quest for the meaning of life; dedicate yourself to a quest for nightingales.

It's like the famous apple of Narnia: if you pluck it for yourself, it will bring only dark and terrible things; but if you pluck it for others, it will bring joy beyond imagining. If you seek nature for what it can do to improve your life, you will have comparatively limited success. But if you pursue nature out of love, you will find a great deal more than you bargained for.

There's a scene in one of the old James Bond films (*Diamonds Are Forever* with Sean Connery) involving Blofeld's cat, or rather two identical cats. Bond makes an immensely plausible attempt to get the better of his great enemy, picking up one of the cats and chucking it across the room, but alas the plan miscarries. Blofeld gives a patronising sneer: 'Right idea, Mr Bond . . .'

Bond completes the sentence: 'But wrong pussy.'

Time and again as I turn left towards the water meadows, I pass people hammering past in the opposite direction, running hard and doing their damnedest to just run along. I give a cheery greeting and it's ignored. They can't hear me. They've gone jogging and they've filled their ears and their souls with music to make up for any shortfall in the wild world. Right idea to get out into the countryside, right idea to get those muscles moving, right idea to improve that cardiovascular fitness – but wrong to stop your ears from the wild world you're moving through. Right idea, but wrong pussy.

It's ten times worse when I visit my father in London. The towpath by the river is thronged with joggers and also with cyclists determined to be as bullying to walkers as the cars and lorries are to them on the main road – and all with their ears blocked, as if the sounds of nature were a force for terminal harm. Let's have a bit of nature – but nature on our own terms. It may be thrilling to hear the calls of the first terns that hit the river in late spring, but what really matters is my time for my run. I sometimes feel as if I am walking through a great crowd of Mr Joneses in the Bob Dylan song: 'You try so hard but you don't understand . . . do you, Mr Jones?'

The spell of this chapter will not bring you every answer to the conundrum of physical improvement. You won't be able to look at the read-out on the device strapped to your wrist and learn about the number of calories you have burned off or how much the cubic capacity of your lungs has improved or how much longer you will now live as a result of your punitive exertions. But it will help you to love nature a little more: and, as a result, you will feel just a little loved back. This may even help you to live longer. It will most certainly help you to get more of a kick out of life in the good times, and give you a certain small consolation when things are less good.

So get outside. Get outside somewhere nice and walk and look and listen. When you set off, don't ask yourself how much exercise do I need – ask instead what bit of nature you would like to pass. To be, however briefly, at one with.

The secret is all in the subtle adjustment of priorities. Climb the hill not because you need the workout but because you might see a buzzard. Take the loop through the park not for the sake of your gluteal muscles but because there are sometimes redwings on the grass in winter and blackbirds singing in spring. Once you start to think like this, everything changes.

You are still taking exercise, but you are no longer Taking Exercise. *You* are no longer the priority. You are an innocent victim, caught in the crossfire of nature.

My own first rule is to bring binoculars. Even if you never raise them to your eyes, they are a token of your sincerity: a sort of personal guarantee that the wild world is your priority. The bins set the tone.

And as you get into your stride, you will begin to invoke some of the other spells in this book. Listening, for a start: I'd reckon that for every species of bird I recognise by sight, I find twice as many by ear. Even if your skills are limited, the thrill of listening out for one or two – or even the magical five I have already offered – is considerable. The jogger who hurries past you, irritated as you pause in the path to look up into the tree at the sound of a great-spotted woodpecker, is moving to the sound of 'Hit Me With Your Rhythm Stick'. You are moving to the rhythm of the planet and the music of spheres: the robin in winter, his sweet lone voice, or the blackbird in May telling you that there are pleasures beyond sex and drugs and rock and roll.

Your peripheral vision will be bringing you information from all around you. Your magic trousers – or the magic plastic bag in your pocket – will bring you a dry place to sit every time you want to stop and contemplate great matters. In the warmer months you will see not butterflies but peacocks and brimstones. The cubic capacity of your lungs may or may not be expanding, but the cubic capacity of your heart and mind and soul is now greater than you ever dreamed it could be.

Walking, pausing, resting, listening, responding, seeing things from the tail of your eye and allowing them into your experience: all these things are making you wilder, and as they make you wilder, they make you richer.

You know, I always thought – thought for years – that great-spotted woodpeckers were comparatively unusual birds: birds that you boasted about if you were ever lucky enough to encounter one. Then I discovered the spell that brings them to life. *Pik! Pik!* Look up the sound at once, look it up on your brand-new birdsong app, or just search for 'RSPB great spotted woodpecker'. The brief account of the species includes an example of that sharp, far-carrying call. Get it lodged in your brain – make a woodpecker-shaped hole in there – and next time you walk across an urban parkland or anywhere there are mature deciduous trees, chances are you will hear that call. Look up: you may see a shape in the upper branches, especially if it's winter and the branches are bare. Or you may see the bird launch itself between one stand of trees and another: the flight steeply undulating, crossing the open space in a series of waves. A bird you thought rare is now with you always: that is the joy of the ordinary, and it is part of every walk you ever take, if you have eyes and ears open to the joy of it.

Your eyes are better now, or rather your brain is. You respond more quickly to the things around you. And when you see a bird moving with that flashy power and speed, you know – no, it's not a pigeon, though it looked like one for an instant, its flight is too glidy, its wings are too sharp, shearing the air like a pair of scissors. It's a falcon, and in all probability a kestrel, the familiar bird of the motorway verges. Hopkins's windhover rebuffing the big wind . . . or is it?

You have learned not to make easy assumptions and, besides, you have your bins with you. You raise them to your eyes because, well, it's always nice to see a kestrel in action and – well, maybe it isn't a kestrel.

And it isn't. It's too big, too burly, too powerful, too damn fast – and there in your bins it shows not a bricky-orangey-red but apparently black and white and with a huge cad's moustache. It's a peregrine falcon: over the River Thames, over Norwich city centre, over the marsh near my house: a wonderful bird that's only available to those who look up. And that is the joy of the extraordinary, and it's not part of every walk you ever take, but it's part of a lot more walks than you ever thought it could be.

Here are the great gifts of nature – given more freely to those who ask than to those who demand. Ask, then, and receive both humbly and gratefully.

21

How to Create Life

The King took the bucket in both hands, raised it to his lips, sipped, then drank deeply and raised his head. His face was changed. Not only his eyes but everything about him seemed to be brighter.

'Yes,' he said, 'it is sweet. That's real water, that. I'm not sure that it isn't going to kill me. But it is the death I would have chosen – if I'd known about it till now.'

– *The Voyage of the Dawn Treader*, C. S. Lewis

Sun plus water equals life. That's a rough-and-ready equation that holds good in most practical circumstances. You could accurately point out that there are other ways of sustaining life, but on the other hand, the possibilities of constructing a hydrothermal vent in your back garden are fairly remote. (In parts of the ocean where tectonic plates collide, there are life systems powered by geological activity, rather than the energy of the sun. These are fascinating places, and they have made necessary all kinds of adjustments in our understanding of the way life works. But it remains true that so far as our day-to-day lives are concerned, the sun-and-water notion is good enough.)

It follows, then, that where you find fresh water, you will generally find life. We look for water all the time without knowing it: our eyes and our wild minds thirst for the stuff. Our idea of a nice view tends to involve places where we can see for a fair distance, preferably with mature trees in sight, and a decent expanse of water. As usual, this is a taste that goes back to our ancient and atavistic selves: in such a landscape we get fair notice of any danger that may come our way, we can often see large grazing mammals, wild or tame or even halfway between the two, promising future feasts, while the water offers freedom from thirst and life into the morrow for us and for everything else in the landscape. This ideal vista is the place where we feel most safe: a landscape in which we feel we have a future.

In looking for what humans want it's always instructive to see what rich people actually get: and so often, that's a large country house set in a park full of stout trees with an open sward grazed by deer and a lake running right through the middle. That's the human idea of the good life, and it's why rich people have created it again and again. It just happens to be a reasonable imitation of the wooded savannahs of Africa: the places where we first learned to be human. That atavistic part of our minds is what we consult when we look for adventure, for safety and for our deepest understanding of life.

Look out of your window, then. And if the possibilities of establishing a deer park and a square mile of open lake are beyond you, don't despair. The fact is that even the smallest acreage of water will improve your view. It will improve it for you, and it will improve it for the living things you share it with.

It's remarkable how little water you need to effect a change. Access to open fresh water is something very many living things require on a daily basis, and this is not as easy to find as you might think in a landscape built by and for humans. In towns and cities it can be downright elusive. So if you can supply even a little fresh water to any bit of landscape under your control, you will be doing a service to many creatures. You will also be performing a double service to yourself, for you will be able to watch that water and keep an eye on the comings and goings.

A garden pond is one of those heartily recommended things, but establishing one always sounds a little intimidating. It seems to be a major project, involving serious disruption to your way of life. The answer is not to be intimidated: if you just put a washing-up bowl into the garden and fill it up, you will be further on than you were before.

Many years ago I lived on the edge of London, just within the M25. And I was filled with a mania for a garden pond. The fact that there was no room in the garden for such a thing was no deterrent at all. Eventually we found a place in which we managed to insert a pond two or three feet square. It was dug and lined and planted with a couple of off-the-shelf water plants and that was it. The cost was tiny, and so was the pond. It looked like a puddle with a little attitude. It needed constant topping up in hot weather, but it was so small that was never much of a chore: just a splash from a bucket filled at the kitchen sink. I liked the pond because it showed that we were trying to do the right thing: and if that was all it ever did, it would have been enough. But things got better.

I remember a moment of perfection: the first time I saw the pond really in use. I was looking out of the window into the garden and it seemed that someone out there was systematically destroying the pond just to spite me. There was water everywhere: being hurled skywards as if the further it travelled the better it was for all concerned. I moved to another window, where I could see the pond properly: and right in the middle there was a cock blackbird taking a bath with reckless enthusiasm. It was a performance of complete commitment. Cleanliness and feather maintenance are vital to every flying bird: the mechanism for keeping airborne is delicate and needs to be in the best possible order. And there was a blackbird getting on with the job, and all thanks to our pond. I am keeping that blackbird going, I thought. This garden is good for blackbird-kind. Because of this garden, his life is a little bit better than it was before.

I had no great hope of any more specialised aquatic life coming to the pool, but water is the most extraordinary stuff and those that need it most can find it from incomprehensible distances. I was walking past the little pond and making a sort of snap inspection as I did so – and then I performed the most massive and hammy double-take. What's that in the pond?

I dropped to my knees and saw that, impossibly, lurking comfortably beneath its surface were two or three small dragons. Newts, of course. I was astonished and gratified in equal measure. Naturally I had read that newts are supposed to turn up in garden ponds, but I had never for a second believed they would actually come. That had seemed one of those fantasies that wildlife people indulge in.

But there, at the bottom of our tiny pond – a pond hardly worth the name – were three common newts. The nearest pond I knew of was half a mile away: these newts had braved the gardens, the vegetation and the dew-soaked pathways to make it here, running the gauntlet of cats and dogs and humans, and there they were in this tiny refuge, looking as if they had been sent by witchcraft.

An Englishman's home is his castle, but a castle is nothing without a moat. A little water defines the place, makes sense of it, adds a richness to everything all around. A pond is the centre of the stage: the garden's focal point, drawing the eye, bringing in many other living things. More than that: it's a statement of what a garden is for. Once you have a pond, it's a sign that you have abnegated some of your sovereignty. The garden is not there to show off your gardening skills: it's there to gratify a far deeper level of human pleasure. The garden is not for humans first and foremost and always and only: it's for every living thing that might find the place of use. That shift of attitude changes your understanding of your garden – and then of all other kinds of open space. If a space is good only for one species among the ten million or so that make up the animal kingdom, it's not really much cop, is it? It's narrow-casting to a demented degree. And if your own garden is about species beyond our own, then so is the rest of the world.

Should I talk about feeding the birds? If you're reading this book it's highly likely you already put food out for birds. It's a good spell to bring wildlife closer into your life, that's for sure, but it's a spell widely known to all, and available in every supermarket and garden centre. Few wildlife spells are as charming or as easy as the summoning of a party of long-tailed tits down from the treetops to your own windows: tastefully decked out in grey and pink and si-si-si-inging to each other all the time, because keeping in touch with each other is almost as important to them as the next food – and the next drink.

You can also put out food to attract mammals, though not invariably the ones you want in your garden. Dog food and non-fishy cat food can bring in foxes and hedgehogs... though the biggest favour you can do for hedgehogs is to make discreet holes in your fences, to allow them to roam from garden to garden.

Water, then, is at the heart of it. Add water – fresh, accessible water – to any open space that you control and you double its value to the rest of the world. But the real spell is what this water does to you. You bring more life to your garden, sure, but that's because you have begun to abandon the notion that this space is your space and that humans have sole rights over it. Your garden is a good deal wilder – and so, dear reader, are you.

22

Fantastic Beasts and Where to Find Them

The Road goes ever on and on
Down from the door where it began.
Now far ahead the Road has gone,
And I must follow, if I can.

– *Lord of the Rings*, J. R. R. Tolkien

Bilbo Baggins had set his heart against adventures. He had no intention whatsoever of leaving the Shire and seeking the unknown. Such things were for other people. He had no idea that adventure was what he craved in his heart: he had no idea that adventure would fulfil him and make him complete. Above all, he had no idea that an adventure can begin with a single step outside your own front door.

It is the same for us all. There are adventures just beyond the door for anyone who chooses to take the path that's offered. Bilbo came across a dragon; what will you find? The same thing, I expect – that is to say, more than you bargained for.

We modern humans seem to have decided that the most thrilling forms of wildlife are beyond our scope: for the expert, for the specialist. We have no hope of encountering such exoticisms for ourselves: instead we have television, and we have our imaginations. It seldom crosses our minds that an awful lot of these wonders are available to everyone – and that some of them are quite dismayingly easy.

'I've always longed to see a puffin.' How many times have I heard that plaintive remark? It's generally expressed in the sort of tone you'd expect if the person had always longed to see an angel – as if an expedition to see a puffin would require months of planning, specialist equipment of the most colossally expensive kind, along with the luck of the man who broke the bank at Monte Carlo. The fact of the matter is that all you need is a train ticket, or the means to make a car journey.

We are the tiniest bit reluctant to make a journey for the sake of the wild world. Perhaps once again that's something atavistic in us: our ancient minds still assume that wild creatures are all around us in vast and threatening numbers. Alas, that's no longer the case. If we want to see wild things, we need to travel a little. But never as far or as desperately as we might think.

It's a shocking fact that many of the most amazing wild creatures on the planet are out there more or less waiting for us to turn up. It's as if you could purchase a day-return to a dragon's lair, or to the nest of a hippogriff, or even to the vaults of heaven where you can find yourself surrounded by squadrons of angels.

So I am going to offer you six trips to wonder: six adventures that will bring you to magical beasts and allow you to spend a little time in intimacy with them. These great sights are available to you and are not even slightly difficult – and, having experienced them, you will be aware, as never before, that such matters are part of your concern. And they're all in Britain. So let's start off with those puffins.

1. Puffins

The easiest way to see puffins is to travel to Bempton Cliffs in Yorkshire. Good views are guaranteed: all you have to do is turn up. You don't even need to get on a boat. There is a visitor centre run by the RSPB and it's the gateway to a quite overwhelming experience. The sight is something, but the sound is something greater – and greater than either is the smell. Here on the cliffs, in late spring and early summer and in hundreds and thousands are the seabirds, and the guillemots and razorbills dominate. They lay eggs on the ledges of the cliff and raise their young in this tumultuous vertical city. On the clifftop the puffins do their stuff in burrows – and you are shocked that so unlikely a bird should look so much like its own illustrations. They are smaller than you expect, and every bit as charming and as comic. You never believed that puffins were really real, and now they're all around you.

2. Dolphins

An understanding of place is crucial to understanding the wild world. Food, shelter, safety: all of these things matter. And though dolphins are fast and mobile, adaptable and intelligent, they have a special taste for the fish-rich and sheltered waters of Cardigan Bay: the patch of water that lies between the two great arms of Wales.

Bang in the mid-point of that great bay is the small seaside town of New Quay and it's one of the best places in the country to see dolphins. There's a permanent population that lives out in the bay and they are relaxed and confident around humans and their boats. I have seen them in the harbour itself on more than one occasion: cruising about while people eat ice creams. There are also sea trips, going out every day during the warmer months. It's always seemed to me the most amazing thing of all: that from the vastness of the sea there emerges, impossibly, a creature sleek and gymnastic and as much at home in the water as you are in your own sitting room – and yet breathing just the same as you and me. A single second with a leaping dolphin is enough to make a believer of anyone.

3. Eagles

The island of Mull offers two species of eagle and, as if that wasn't enough, it also distils two different malt whiskies. I was able to sample all four, without very much trouble but with the most immense joy. Getting there is simplicity itself: a short ferry crossing followed by that thrilling sense of isolation that comes with a visit to even the most accessible of islands. The white-tailed eagle went extinct during the great Victorian persecution. They were first reintroduced on the island of Rum in 1975 and have since prospered: and now Mull offers the best opportunity to see them. Eagles have become part of the economy of the island: there are boat trips and other opportunities to look for them. Away from the sea cliffs you can also find golden eagles. And in the great impossible size of both these monumental species – in the glorious, gliding confidence of them – you will find a wild response within you that will take your breath away. You didn't know the world was quite as wild as that. You didn't know you were that wild yourself.

4. Nightingales and avocets

There are certain places that seem to have been let off all the normal rules. Here the animals seem tamer, closer, more confiding; the rare ones seem to have become common, and human beings seem to have been granted a level of trust that they never have in the usual run of places. Minsmere is like that: Minsmere, the great RSPB reserve in Suffolk. Have no fear in presenting yourself in its car park: this is a place that welcomes everyone. Those who know next to nothing will be helped and the daily wonders will be made available to them. From March onward the area of open water – the Scrape – will be full of elegant avocets, birds that went extinct as breeding birds in this country, made a staggering return after the Second World War and are now the RSPB logo. In spring – the first week of May is best – you will also have a chance of listening to the wondrously cacophonous sound of nightingales: staff will tell you the best place to listen. Sit quietly by the reed beds, again in spring, and you might hear the mysterious, far-carrying boom of a bittern. Minsmere astounds: routinely overwhelms. No one goes to Minsmere for the first time and leaves unchanged, so don't say you haven't been warned.

5. Gannets

Bass Rock is the single most amazing place I have visited in Britain. I always thought that it must lie miles out to sea: that paying a visit was a desperate venture suitable only for the most intrepid of seafarers with cast-iron stomachs. Turns out you can see it from the train. And the golf course. Bass Rock lies just off the town of North Berwick, a forty-minute train journey from Edinburgh. It stands up boldly from the sea, shaped like a wedding cake and shining white: not with guano but with the birds themselves. More than 150,000 gannets live on this rock, and just to stand on the shore marvelling is a fine experience. But, again, you can take a boat trip for a few pounds and get right among them: thousands and thousands of vast birds, all with six-foot wingspans, pure white apart from the stiff black wing-tips and the yellow heads. You will see one or two fishing, maybe many: crashing headfirst into the water from a height of 100 feet. To visit Bass Rock is to accept that you had no idea so many marvels were available so readily for so little trouble. You know now that you live in a world in which marvels can be had for the asking.

6. Swallowtail

One more wonder, just so that the birds don't have it too much their own way. Go to the Norfolk Broads in the early summer – Strumpshaw Fen, run by the RSPB, or Hickling Broad, run by Norfolk Wildlife Trust – and on a fine sunny day, even better if it's pretty still, you have an excellent chance of finding swallowtail. Look for a great yellow splash among the reeds: if it's obviously too big to be a butterfly, then you've found it – and it really is a butterfly. There are boardwalks to help you, and on the right sort of day there will be other wildlifers happy to point you in the right direction. Or you can hire an electric boat on Hickling Broad – quite a lot of it is navigable – and as the engine murmurs almost silently you can look for those impossible flashes of colour.

* * *

Let's leave it there. Plenty more wonders, plenty of them easy enough to see – yes, and to hear and to smell. Once you have made a journey for the specific purpose of seeking the wild, you will have done something to your mind. It's that little bit wilder. You are now more likely to see wonders than you ever were before. You've seen things that amaze, and you'll be less surprised – and more ready – the next time you're in the right place for wonders. And here's a rum thing: the more you see, the less surprised you get – but the more amazed.

That first naive shout of delight at an encounter with something impossible is not a peak experience, though it feels so at the time. The more you see, the deeper and richer your sense of delight will become. The first kiss is not the best kiss: it's merely the most exciting. The most recent kiss is always the best of all. The greatest kind of amazement comes from familiarity. And the wilder you are, the more amazing life is.

23

Deep Magic From Before the Dawn of Time

'It means,' said Aslan, 'that though the Witch knew the Deep Magic, there is a magic deeper still which she did not know. Her knowledge goes back only to the dawn of Time.'

– *The Lion, the Witch and the Wardrobe*, C. S. Lewis

So let's say you have read all the spells in this book and have put a good few of them into practice. As a result you are a better naturalist than you were before. You will find and recognise and name things that were once beyond your scope. You have improved your knowledge and your skills, but even as you do so you become aware that this is the lesser half of the magic I have been trying to show you.

You are seeing more. You are noticing stuff that you'd never have noticed before – so much so that you wonder if all these wild creatures were even there at all, in the distant days before you began to learn a few of these rewilding spells. It's as if something in you – something new and wonderful – has called them into being. You know that this is not magic, in the strict sense of the term, but you also know that it's as close as you're ever going to get.

You can savour this magic: recognising the song of a song thrush, finding a slow worm, noticing that a deer has crossed your path, seeing a small tortoiseshell butterfly flying past intent on its own urgent business, noticing a robin drinking from the plastic pond you installed yourself. All these things enrich you, but even as you enjoy them you are beginning to be aware of a deeper truth. These things you sought are only the beginning. They are the symptoms of an altering condition.

You are a better observer, and that is gratifying and exciting. But observation – when it involves wildlife – is not a one-way street. If you get good at recognising different kinds of aeroplanes, then you are good at recognising different kinds of aeroplanes. There is very little that the aeroplanes can do back. You have the satisfaction of acquiring a mildly unusual skill, and the pleasure of a small involvement in the passing traffic above your head. But that's about the end of it.

Perhaps you thought it was going to be like that with wildlife: that learning to be a better observer would be like collecting the Brooke Bond picture cards of British Wildlife: jolly good fun in a relatively superficial kind of way. But as you try more spells and get better at the ones you've already learned, you begin to understand that you're not going to get this all your own way. By naming the singer in the distant tree you have established a claim on the bird: but the bird has also established a claim on you.

We humans first walked the savannahs millions of years ago, and we did so for millions of years. It wasn't a fly-by-night experience, something we rapidly outgrew in our rush to invent agriculture and civilisation. We were wild for age after age after age: we have only tamed ourselves in the past few thousand years. We have been city-dwellers for just the blinking of an eye. The wild parts of our brain are deeply established and it takes the smallest adjustment to wake them up again.

When our most ancient ancestors were out on the plains of Africa, hunting and gathering in a world that stayed the same from one generation to the next and the next and the next, you can bet all you possess that they were crash-hot at birdsong recognition. They had to be: their lives depended on it. They all knew – men, women and children – that a dry, rattling chirr from deep bush means 'keep away'. It's the call of a flock of oxpeckers, and they'll be feeding while perched on the skin of some large and potentially dangerous mammal. You don't want to walk in unannounced upon a lone buffalo or a hippo on his way back to the river after a night's grazing. You alter your line of march and stay safe for another day.

In the same way, a long, confiding, chattering call is something you change your plans for. This is a greater honeyguide, and if you follow him he will lead you to a source of honey, feeding on the grubs you expose as his reward.

It follows, then, that these long-lost parts of our brains are there and ready to leap into action. They can be reactivated with remarkably little trouble. It's not that the knowledge itself is innate: what's innate is the desire – the need – to know more, to understand more, to establish a closer and more meaningful relationship with the wild world.

That's why the beginnings of understanding are comparatively easy to acquire: our brains are ravenous for a deeper attachment to wild things. Persuading your brain to become a little wilder is like trying to persuade a starving man to take a snack.

And it's at this point that things start to get a little out of hand. You are closer than ever before to the wild world – and then you find yourself asking: the *wild* world? You mean as distinct from the tame world? Are there then two completely different orders of existence: one that involves human beings, and one that involves everything else that lives on this planet? Or is there in fact only one?

You are alone. You have your binoculars, but it's been a quiet day, and perhaps all the better for that. A light shower was falling, but it's easier now and you take it in your stride. You come across a place that's perfect for a sit: in your waterproof trousers – or with your supermarket plastic bag – you have no fears of a soggy bottom. You sit at your ease. Before you is a wild landscape, and it is packed with the doings of many. A bird sings out: it's a song thrush again, repeating its favourite musical ideas before moving on to a new one.

And then silence. Not that it bothers you. You have stepped, if only for a while, beyond the sound of engines and music and human busyness: and what you find there no longer alienates. Why should it? You are part of it; it is part of you. That's not a mystical connection: it's a connection to do with blood and brains and bones and gristle. You, rewilded, sit at your ease in the world, in the great silence. In a little while that thrush will sing out again. And you'll be ready. Now you'll always be ready.